THUNDERSTORMS,

TORNADOES

AND HAIL!

by

Peter R. Chaston

Chaston Scientific, Inc.
P.O. Box 758
Kearney, MO 64060
phone: 816-628-4770
fax: 816-628-9975

Library of Congress Catalog Card Number: 98-94957

ISBN 0-9645172-6-4

cover photo: NWS

Dedication

I dedicate this book to my wife, Mary, and to my daughter, Valerie, who are the joys of my life, and to my second joy: the love of weather. Therefore, I also dedicate this book to everyone who enjoys and is enthralled by the vagaries of weather.

ACKNOWLEDGEMENTS

NOAA = National Oceanic and Atmospheric Administration

NWS = National Weather Service

NHC = National Hurricane Center (Tropical Prediction Center)

Environment Canada

FAA = Federal Aviation Administration

DOA = Department of Agriculture

DOT = Department of Transportation

USAF = United States Air Force

NASA = National Aeronautic and Space Administration

NCAR = National Center for Atmospheric Research

University of Alaska

Virginia Department of Highways

BOOKS WRITTEN OR CO-WRITTEN BY PETER R. CHASTON:

- **WEATHER MAPS, SECOND EDITION - How to Read and Interpret all the Basic Weather Charts (ISBN: 0-9645172-4-8)**
- **TERROR FROM THE SKIES! (ISBN: 0-9645172-1-3)**
- **HURRICANES! (ISBN: 0-9645172-2-1)**
- **JOKES AND PUNS FOR GROAN-UPS, co-authored with James T. Moore (ISBN: 0-9645172-3-X)**
- **WEATHER BASICS, co-authored with Joseph J. Balsama (ISBN: 0-9645172-5-6)**
- **THUNDERSTORMS, TORNADOES AND HAIL (ISBN: 0-9645172-6-4)**

These books are available from select bookstores or directly from Chaston Scientific, Inc.; P.O. Box 758; Kearney, MO 64060; phone: 816-628-4770; fax: 816-628-9975.

THUNDERSTORMS, TORNADOES AND HAIL

TABLE OF CONTENTS

(continued)

TECHNICAL SECTION:

INTRODUCTION

Nature produces some of its beauty in awesome fashion in the products of a thunderstorm, especially in a severe thunderstorm.

Who among us has not marvelled at a display of nearly continuous bolts of lightning in a strong thunderstorm? Many of us must also have been impressed at experiencing large hailstones striking at us, at our homes and on our vehicles. And some of us have had to run for our lives from the most violent of a thunderstorm's creations, a tornado!

Although scientific research into thunderstorms occurred over the centuries, the significant study of thunderstorm development began in earnest in the 1940s, especially after the conclusion of the second world war. By the beginning of the 21st century, meteorologists had learned much about the genesis and life-cycles of thunderstorms, hail and tornadoes, yet what was learned generated even more and new questions which in turn have inspired new research efforts.

From the simple "lightning is electricity" kite experiment of Benjamin Franklin to the sophisticated small-scale computer models developed to study thunderstorms today, the story of this magnificent weather system and its products is a fascinating story indeed.

This book, "THUNDERSTORMS, TORNADOES AND HAIL", describes the current knowledge on these phenomena. The first part of the book is written to be readily understood by anyone with a basic knowledge of weather. The technical aspects of this topic are contained in the chapters at the end of the book.

Our goal, therefore, is to provide you with an information source on thunderstorms, hail, tornadoes and other aspects of thunderstorms. Our laboratory is free, and it is outside. Enjoy!

ABOUT THE AUTHOR

PETER R. ("Pete") CHASTON became fascinated with weather as a young boy. His personal affinity for the science of meteorology began when he experienced a few hurricanes while growing up along the East Coast. He was fascinated by having the eye of a tropical storm named Brenda go right over his home weather station, followed some two months later by Hurricane Donna's 100+ mph winds and driving sheets of horizontal rain. Winter snowstorms and blizzards also thrilled him, and weather grew to be Pete Chaston's main interest.

Having weather as an intense hobby eventually led to a career in meteorology. Pete started reading college texts and everything else he could find on weather through secondary school, and then served as a weather observer in the Air Force for four years, saving money for college.

Pete Chaston received his Bachelor of Science degree in Meteorology and Oceanography from New York University, and later, while a National Weather Service (weather bureau) forecaster, was selected for the weather service Fellowship to graduate school, underwhich he earned his Master of Science degree in Meteorology from the University of Wisconsin. It was at Wisconsin where he met Mary Gabrielski and they married almost two years later.

Pete Chaston served as a National Weather Service meteorologist from 1971 through 1995, afterwhich he took advantage of an early retirement option to found Chaston Scientific, Inc., under whose auspices this book is written.

In the weather service, Pete served at Binghamton, New York and at Hartford, Connecticut before transferring to the forecast office at Pittsburgh, Pennsylvania. He then was the Meteorologist-in-Charge of the National Weather Service Office at Rochester, New York and later became Technical Project Leader for the National Weather Service Training Center in Kansas City, Missouri.

Pete has written several books on meteorology, had a weekly newspaper column on weather, did television and radio weather and numerous talk shows, and is a regular lecturer and speechgiver. He played the role of a meteorologist in the movie, "Water", filmed for the PBS TV network and...for something different... even appeared in a Stephen King movie, "Sometimes They Come Back", and has a popular Kansas City radio program called "The Pete Chaston Doowop Show". He has taught at the State University of New York, the University of Missouri at Kansas City, the University of Kansas at Lawrence, Kansas, William Jewell College at Liberty, Missouri and lectured at other colleges. Pete Chaston has also worked with several grants involving training the nation's earth science teachers in meteorology, and has presented seminars to the National Science Teachers Association and various Academies of Science. He also gives training seminars on weather.

Pete was President of the Kansas City Chapter of the American Meteorological Society for two terms.

Pete Chaston has published scientific research articles in magazines and journals, including the National Weather Digest and Weatherwise. He developed a technique for forecasting heavy snow amounts that is widely used by forecasters nationwide. The technique is called "The Magic Chart" because it is straightforward and easy to use. He also pioneered new operational forecasting procedures now commonplace in contemporary meteorology. Some of the books Pete has written include "WEATHER MAPS - How to Read and Interpret all the Basic Weather Charts", "TERROR FROM THE SKIES!" and "HURRICANES!". With fellow meteorologist Dr. James Moore he co-authored a humorous book entitled, "JOKES AND PUNS FOR GROAN-UPS", and with science educator Joseph Balsama, he co-authored the compendium, "WEATHER BASICS", which is a preferred introduction to meteorology book. Thus, Pete Chaston has varied interests and derives great fun and enjoyment from all of them.

Figure 1-1. The thunderstorm, which should perhaps more appropriately be called the lightning storm. (source: NOAA)

When we look at the early writings and drawings of diverse cultures, we find that one of the aspects of our environment that has always fascinated people is the thunderstorm.

The thunderstorm has a high ranking in the order of weather interests because of its many aspects and sometimes awesome beauty and occasion terror. This storm can produce various forms of lightning, loud thunder, flash flooding torrents of rain, hailstones that can exceed five inches (above 13 centimeters) in diameter, and the deadliest of all the products of a thunderstorm: the tornado. It is to be expected, therefore, that this weather phenomenon would be heavily investigated, written about in science and popular literature and in song, and would captivate the imaginations of scientists and poets alike.

Investigations of thunderstorms and their products go back beyond the 18th century experiment of American patriot and scientist Benjamin Franklin, who discovered that lightning is indeed electricity when he sent a kite into a thunderstorm and felt the electrical discharge when the electricity passed down the kite wire onto the metal key he touched at the base of the wire. The first real investigations by atmospheric scientists (who became known as meteorologist in the 1940s), were conducted in the 1800s.

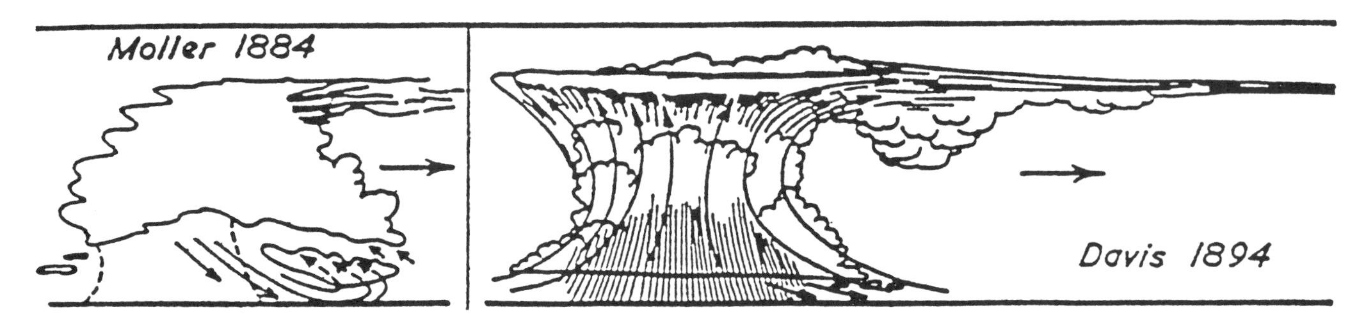

Figure 1-2. Two early meteorologists in the late 1800s theorized the air flow within and in the vicinity of a thunderstorm with these sketches. The arrows indicate what they thought was representative of the average flow of air currents in a typical thunderstorm.

Through the 1900s, it was becoming apparent that rising air currents were needed to form the thunderstorm. Just how these currents of air converged and rose was not well understood, but the realization that the dynamic of thunderstorm development included rising air from below led to the rapid development of theories on the evolution of this storm type.

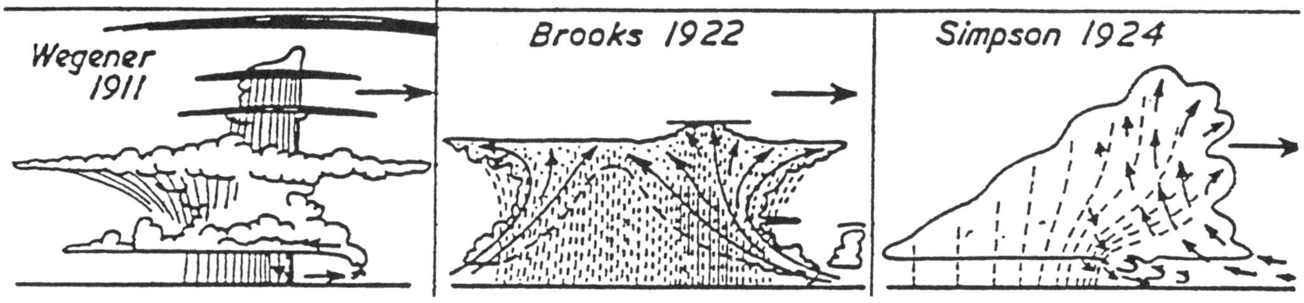

Figure 1-3. Now look at very early 1900s theories of air flow in thunderstorms. By the 1920s, researchers were now concluding that, indeed, there were rising currents, which we now call updrafts, that were required for the development of a thunderstorm.

By the 1930s, the concept was generally accepted that parcels of air in the lower through middle levels of the troposphere in a local area were converging and rising to form the thunderstorm.

Figure 1-4. Here are schematics of thunderstorm air flow from two researchers in the 1930s.

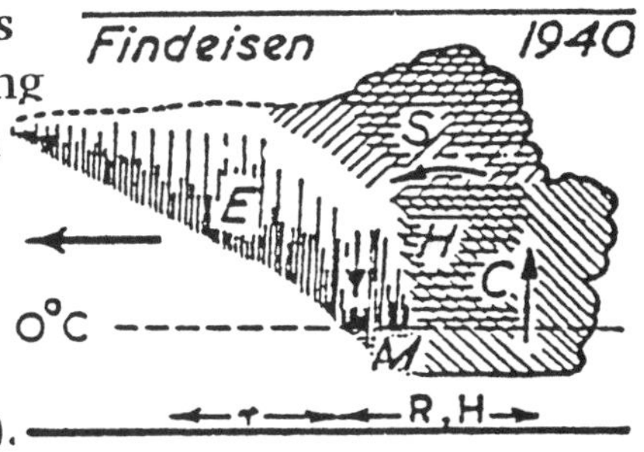

Figure 1-5. In 1940, one weather researcher provided his best guess on a physical model of a thunderstorm, showing an asymmetrical thunderstorm and its areas of evaporation (E in the sketch), condensation (C), sublimation (going directly from water vapor into ice) (S), hail (H), melting (M), hail falling to the ground (also H), light rain (r) ahead of the heavy rain (R). This model suggested that the hail occurred near the thunderstorm updraft (rising air current).

The first real understanding of the dynamics of convection (localized rising air currents made up of buoyant, warm rising parcels of air, and sometimes of buoyant, warm rising layers of air) occurred in the late 1940s when the United States engaged in the first major thunderstorm research project. By this time there were not only surface observations and observations using weather balloons, but also observations from the first weather radars converted from World War II military radars. Moreover, aircraft were flown into and around developing, mature and dying thunderstorms to gather more data on temperatures, humidities, air pressure and wind. For the first time, the existence of updrafts and downdrafts was physically proven through observations.

Two of the researchers, Byers and Braham, published in 1949 the first landmark paper on thunderstorms, describing the three stages of thunderstorm development: cumulus, mature and dissipating stages. They studied thunderstorms in Ohio and Florida. They also discovered that a typical thunderstorm is comprised of several cells, each of which undergoes these three stages of evolution.

These same observational tools were enhanced later by data from newer tools: more and better radars in the 1950s, '60s and 70's, weather satellites which began in 1960, Doppler radars in the 1980s and 1990s and vertical sounders of the atmosphere known as profilers, in the 1980s and 1990s. Models of ordinary and of severe thunderstorms emerged.

As we entered the 21st century, a typical model of a non-severe or ordinary thunderstorm is depicted by the figure below.

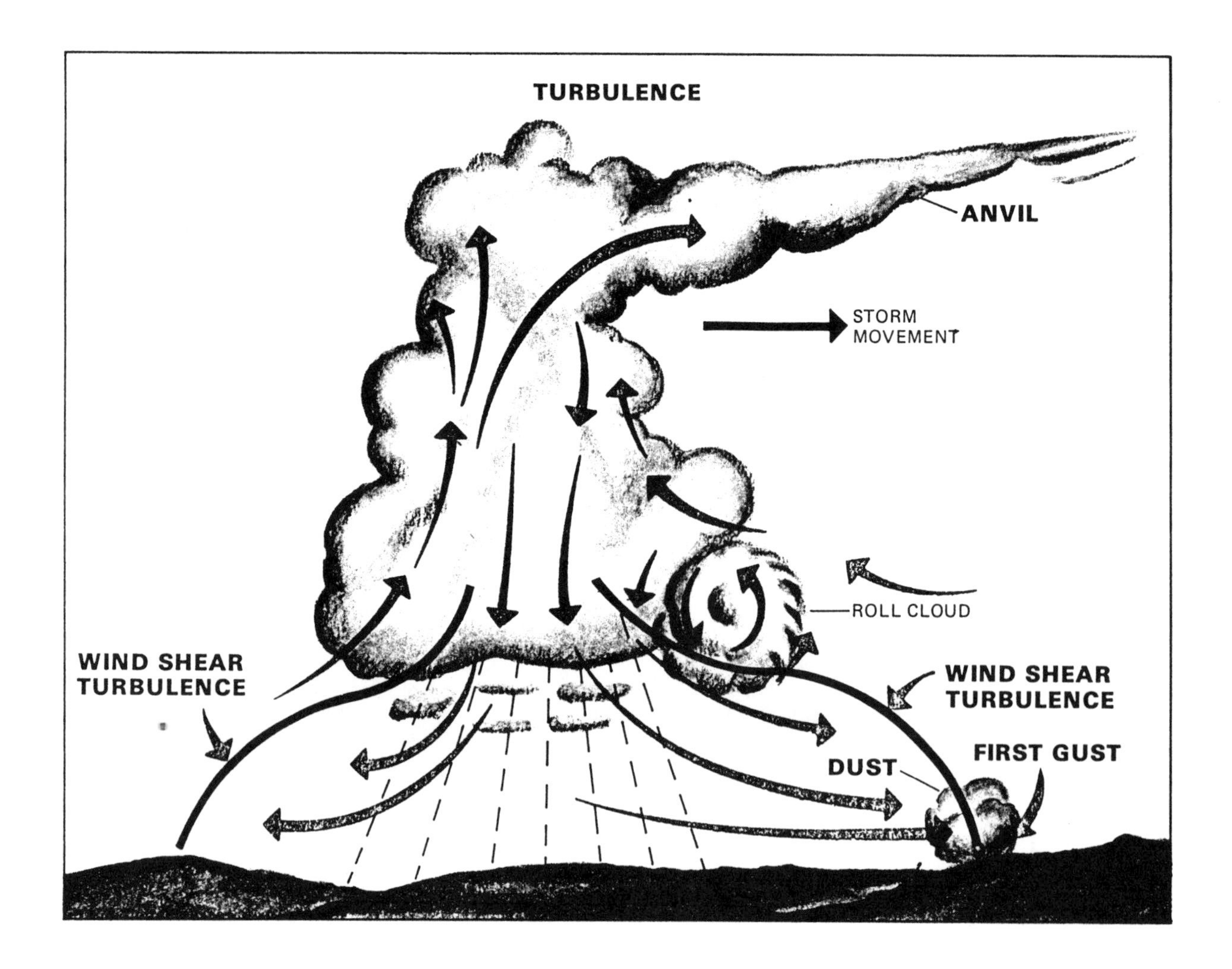

Figure 1-6. A late 20th-early 21st century model of an ordinary, non-severe thunderstorm. (source: FAA)

Now let us start our adventure in exploring thunderstorms, hail and tornadoes by looking at thunderstorm climatology.

Chapter 2. THUNDERSTORM CLIMATOLOGY

First, let us look at some interesting statistics about thunderstorms.

- As you are reading this, there are about 2000 thunderstorms underway around the world.
- Around the world, there are some 45,000 thunderstorms EACH DAY.
- Thus, around the globe there are about 16,000,000 thunderstorms annually.
- In the United States, some 100,000 thunderstorms occur each year.
- During the twentieth century, thousands of Americans were killed by these aspects of thunderstorms: lightning, flash floods and by powerful winds associated with severe thunderstorms, including tornadoes. Very few deaths have been documented due to large hail; one of these was an infant pummelled to death by large hailstones in Fort Collins, Colorado.
- The energy expended by just one typical thunderstorm is enormous. In fact, if the energy from a bunch of these storms could be harnessed, then it could supply the entire planet's energy needs for about one year!

Figure 2-1. These are cumulus clouds, which are clouds of strong vertical development, i.e., formed by rising currents of air. If there is an ample local supply of moisture in the form of water vapor in the air, and the parcels keep rising, then they may grow into thunderstorms.

Thunderstorms occur primarily in the tropics and mid-latitudes, but there is also some thunderstorm activity in polar latitudes. In general, the warmer the climate and weather, the more thunderstorm episodes occur, if adequate moisture is available as water vapor in the air.

Now, let us look at the climatology of thunderstorms in the United States, since the U. S. experiences more thunderstorms than any other geographical region.

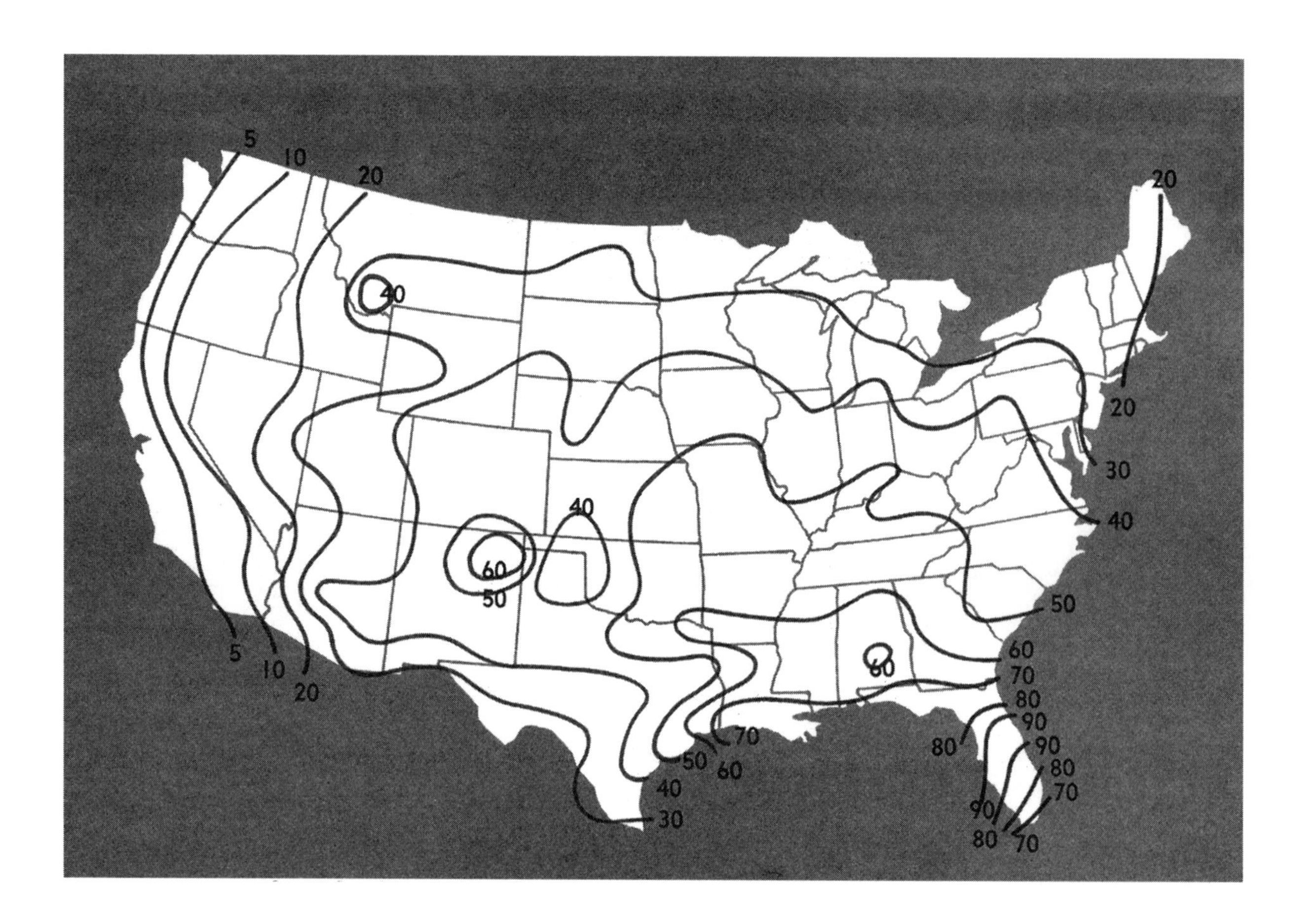

Figure 2-2. Here is the average number of days during the year with thunderstorms in the contiguous 48 states. Alaska and Hawaii each has about ten or fewer thunderstorm days per year. (source: NOAA)

Notice that the warm, moist climate of Florida allows for the most thunderstorm days, whereas the much drier, yet much hotter during the warmer part of the year, climate of southern Arizona has considerably fewer thunderstorm days annually. Notice, too, how few thunderstorms occur each year along the west coast, due in large part to the stabilizing cool marine air that frequents the coastal zone, especially in the lowest layers of the troposphere.

Breaking these down by seasons, we find that most thunderstorms occur in the spring and summer, and the fewest occur in the winter. Let us look at this climatology season by season.

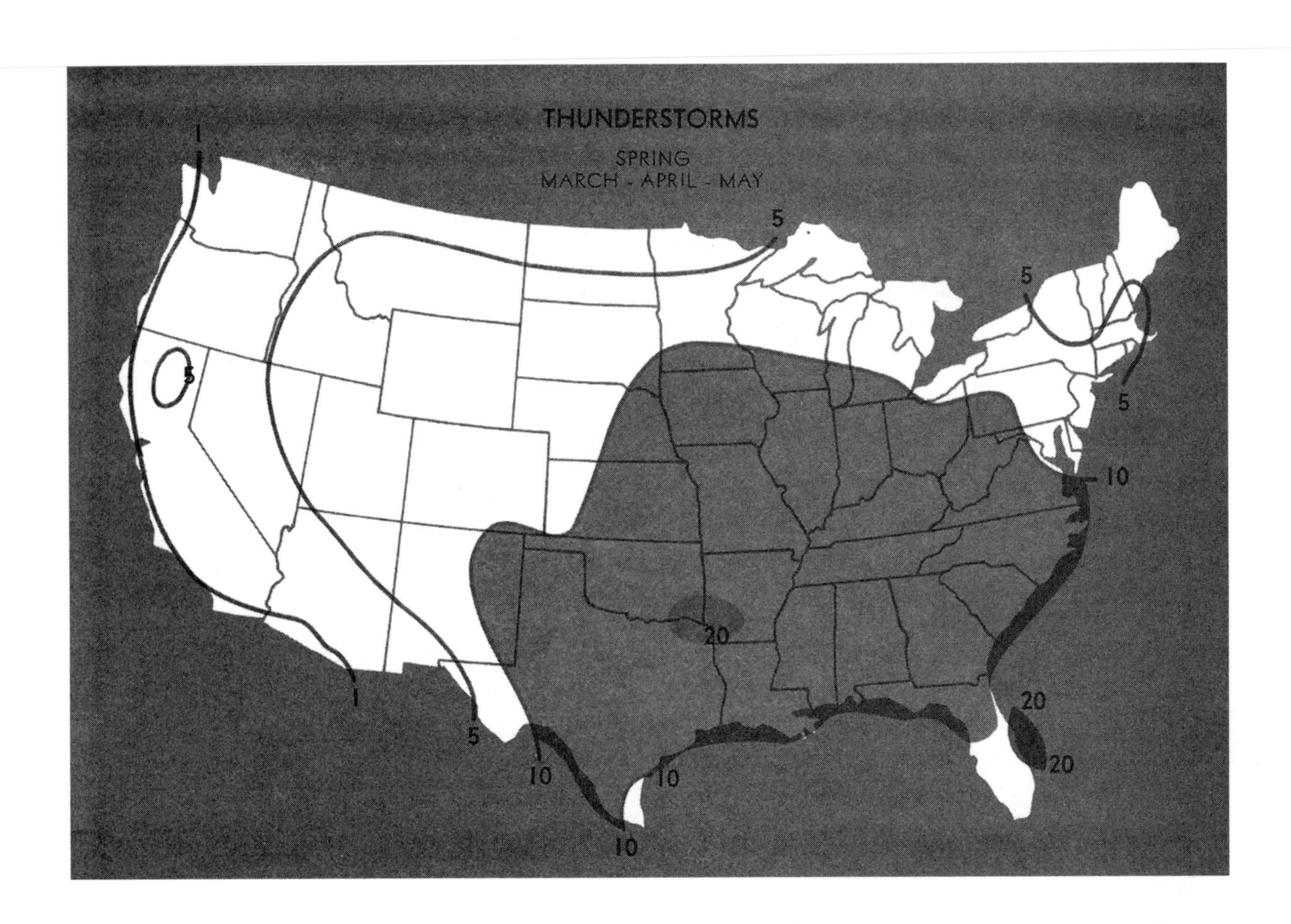

Figure 2-3. The average number of days per year with thunderstorms during the SPRING in the lower 48 states. **(source: NOAA)**

Notice that the southeastern United States, mid-West and Central Plains states typically are the most active regions for thunderstorms during the springtime. Note that the meteorological seasons are not the same as the astronomical seasons. In the Northern Hemisphere, meteorological spring is March, April and May, summer is June, July and August, fall or autumn is September, October and November and winter is December, January and February. This coincides with the hottest temperatures in most places in the Northern Hemisphere occurring during the period June through August, and the coldest temperatures occurring December through February. The other months are the transitional periods.

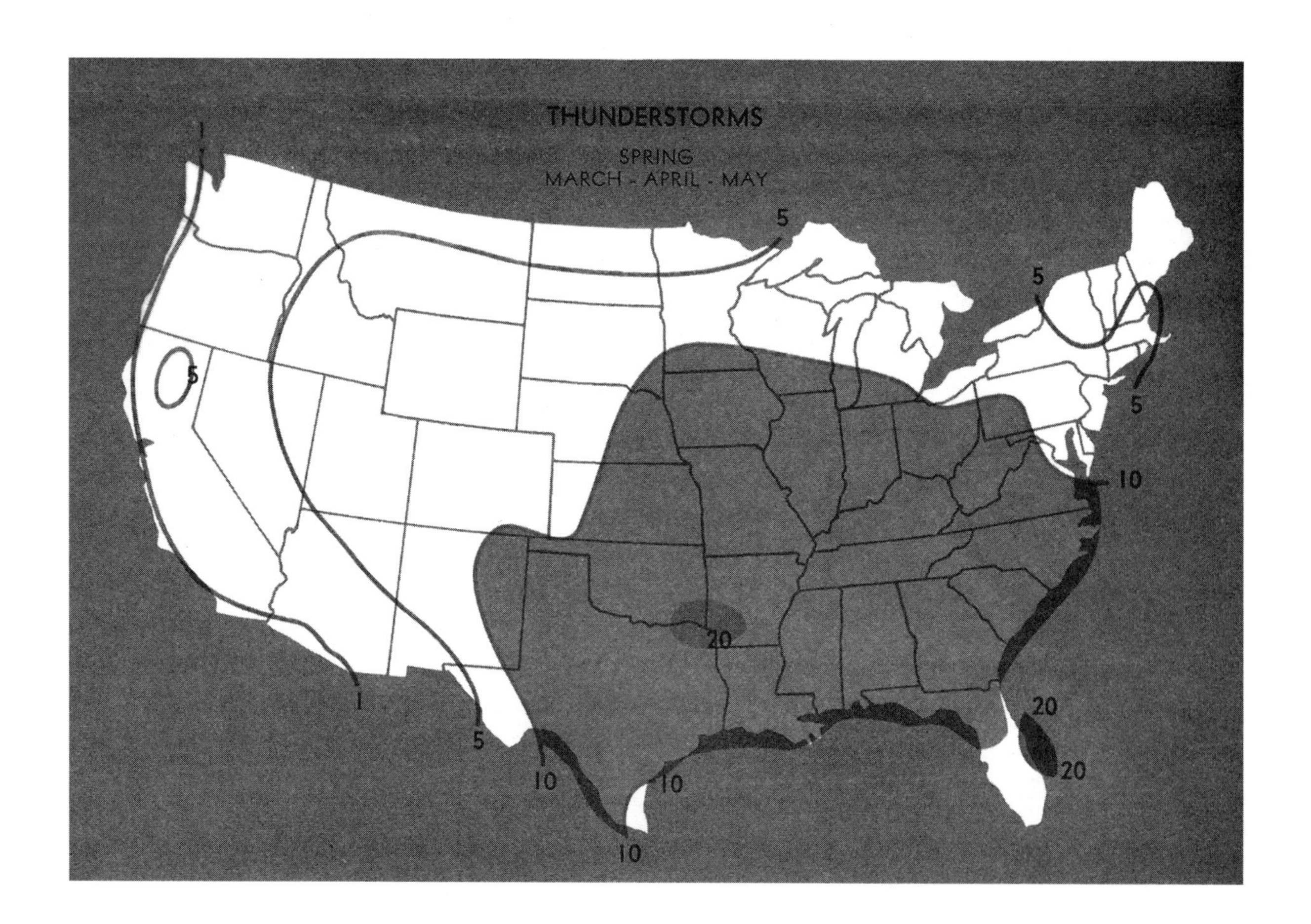

Figure 2-4. The average number of days per year with thunderstorms during the SUMMER in the lower 48 states. (source: NOAA)

Notice how virtually the entire country experiences a significant increase in thunderstorm activity during the summer months. Hotter temperatures allow the atmosphere to hold much more water vapor than during the other, cooler, seasons. Most of the flash flooding thunderstorms occur during this period.

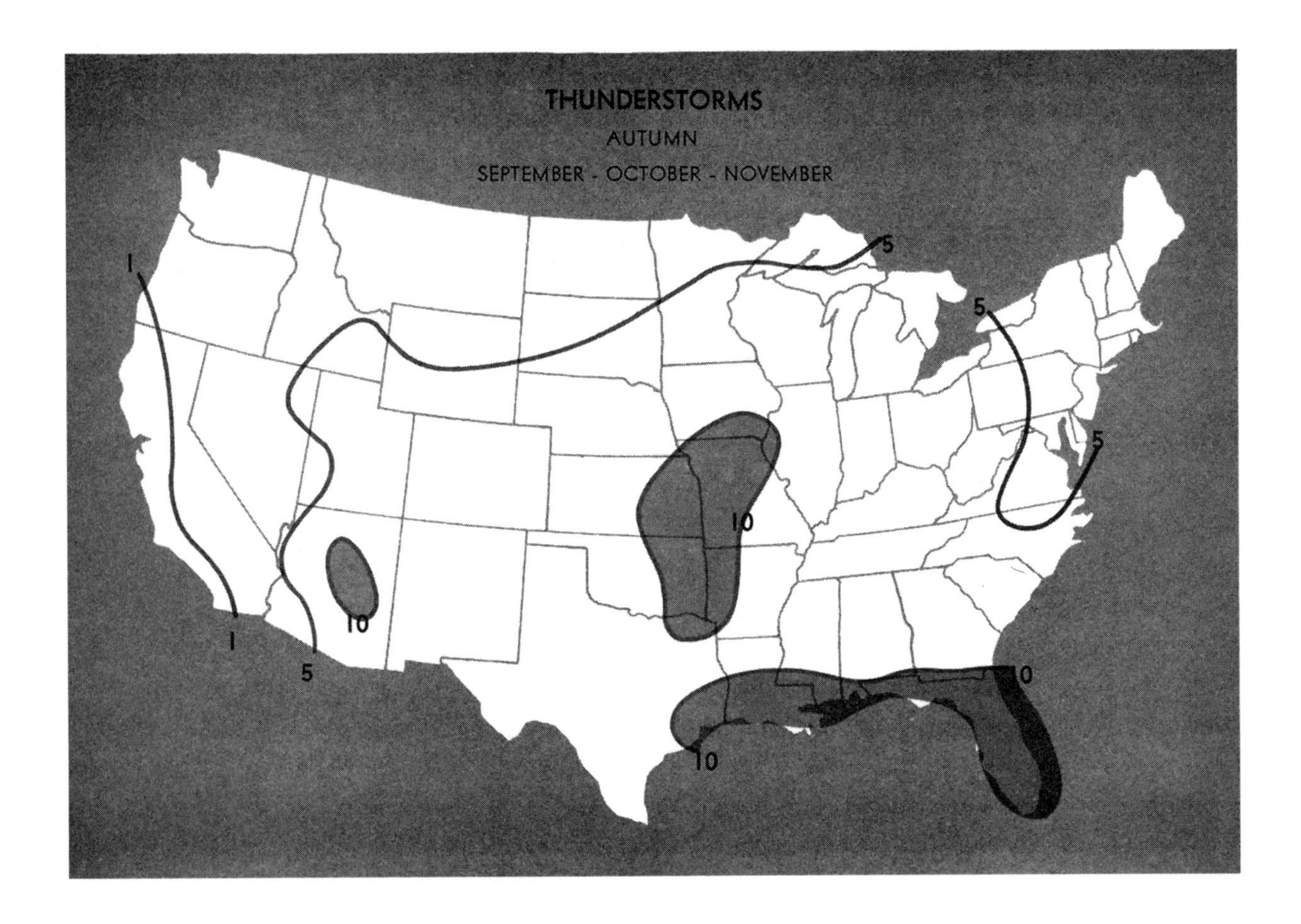

Figure 2-5. The average number of days per year with thunderstorms during the FALL in the lower 48 states. (source: NOAA)

Notice the dramatic drop in thunderstorm days across most of the nation as the climate transitions into the colder months.

Compare the spring activity of figure 2-3 with this fall activity. Notice that there is considerably more area receiving somewhat more storms in the spring than in the fall. This is in large part due to the temperature lag in the middle and upper troposphere. While the lower levels warm up more readily, the levels higher up in the troposphere take longer to warm up. This means that during in spring in most places over land, there is a significant temperature drop as we ascend. When the environment has much colder air over low-level warm air, the local environment is said to be unstable. This instability is one of four necessary requirements for the generation of thunderstorms, as we shall study in later chapters.

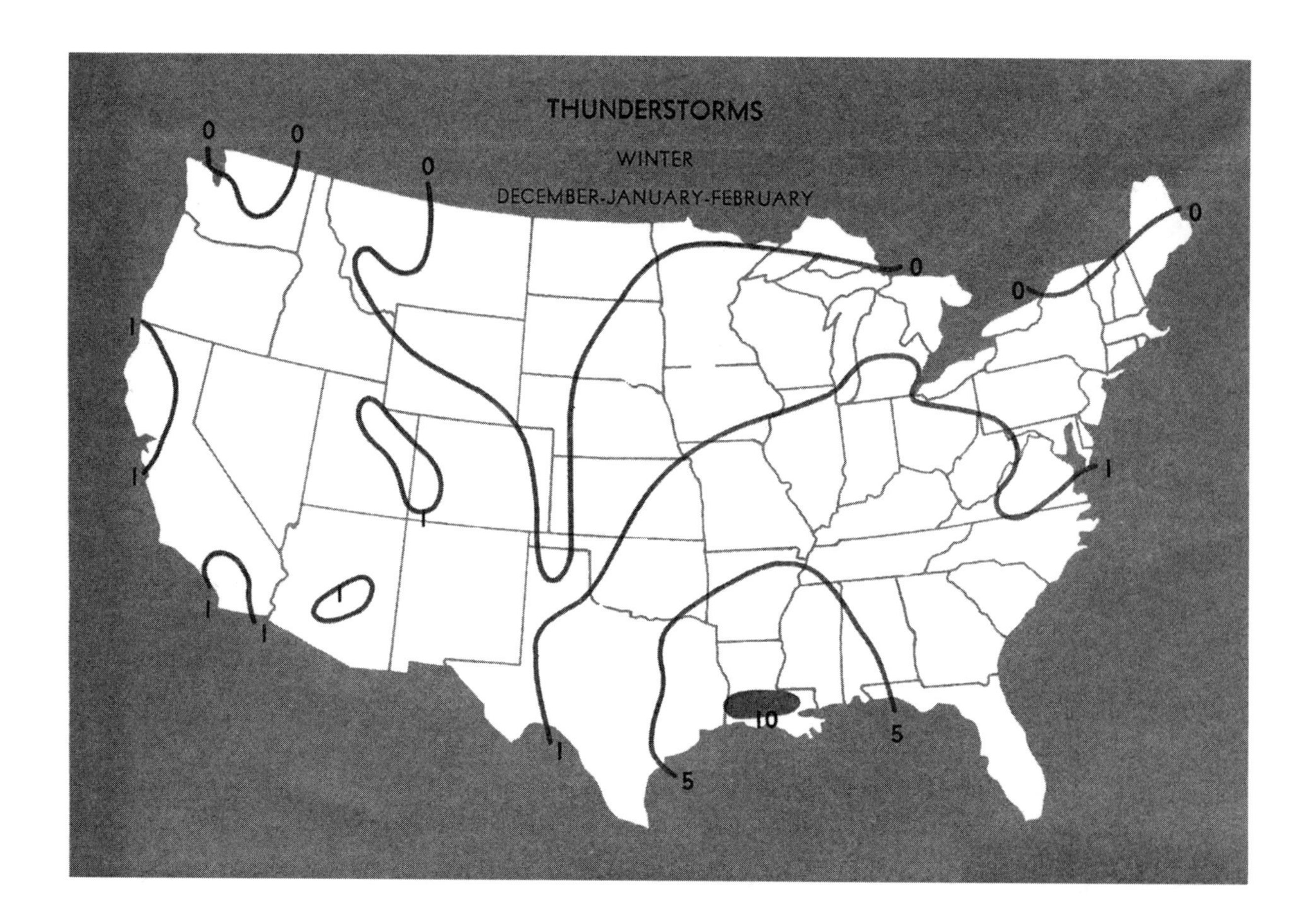

Figure 2-6. The average number of days per year with thunderstorms during the WINTER in the lower 48 states. (source: NOAA)

During the wintertime, the most significant thunderstorm region is typically confined to the central Gulf of Mexico states. Occasionally, thundersnow occurs, imbedded in major winter storms. Also, from chiefly November through March, some snowsqualls from over the Great Lakes grow into thundersnowsqualls as the convection grows to a depth of some 25,000 feet but does so at an angle, which is called SLANTWISE CONVECTION, rather than the vertical or nearly vertical convection of most thunderstorms.

Now let us look at the time of day with the highest frequency of thunderstorms during the most active season for thunderstorms in the 48 contiguous United States.

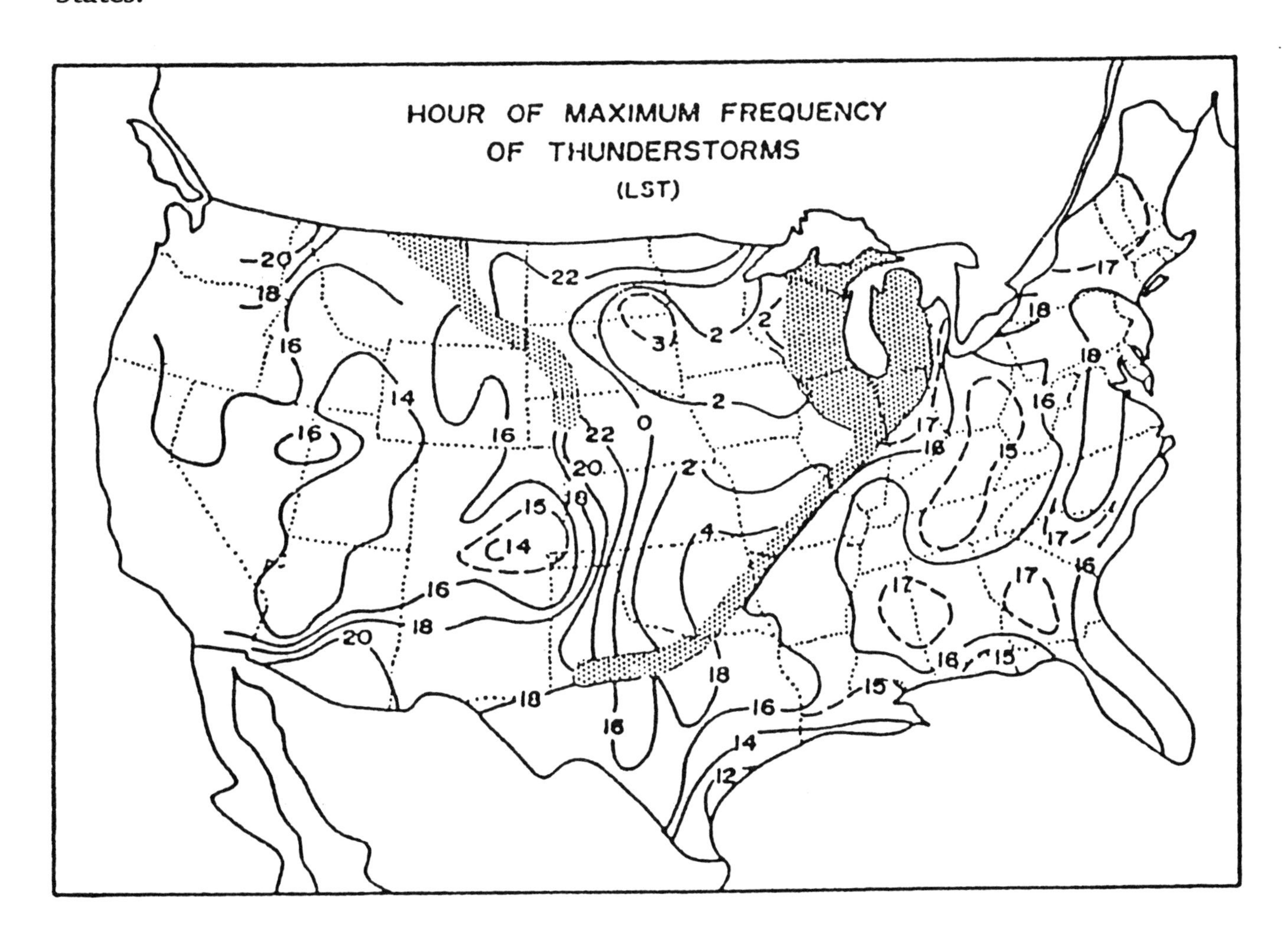

Figure 2-7. **The hour, in Local Standard Time (LST) of the maximum frequency of thunderstorms during the summer, June through August, in the 48 contiguous United States. 0 is midnight, 6 is 6 a.m., 12 is noon, 18 is 6 p.m., etc.**

One feature stands out: in the central plains states, the hours of maximum frequency are very late at night...from midnight to 4 a.m., local standard time (add one hour for daylight time). The dynamics for this nocturnal thunderstorm activity will be discussed later in this book. Elsewhere, and typically, most summer thunderstorms occur late in the afternoon. The heating of the ground by the summer sun plays a major role in the generation of these storms.

Chapter 3. STABILITY AND INSTABILITY

As part of the understanding of thunderstorm development, it is necessary to know what is meant by a stable and an unstable atmosphere.

CONVECTION, or the rising of air parcels that leads to the development of cumuloform (puffy, cauliflower-looking) clouds (chiefly, cumulus, cumulonimbus, altocumulus and cirrocumulus clouds), occurs only when favorable atmospheric conditions (explained in chapter 4) exist simultaneously in the same region. One of these conditions is an unstable atmosphere.

Consider a parcel of air. In meteorology, a parcel is typically from a cubic foot to a cubic yard of air, or, more commonly, from a cubic foot to about a cubic meter of air. If a parcel of air is pushed to one side or another and, after the force that did the pushing stops pushing, the parcel immediately returns to its original location, then the environment of the air parcel is said to be stable in the horizontal direction. (These stability examples can also be done with parcels of other fluids, e.g., water.) Now consider a parcel pushed either upwards or downwards. Similarly to the horizontal example, if the air parcel immediately returns to its origin after the pushing force stops pushing, then the environment of the air parcel is said to be stable in the vertical direction.

Now, suppose the air parcel, once pushed either horizontal or vertically or both, has the application of the pushing force stopped, but the parcel still moves in the direction to which it was pushed. Further, suppose the parcel even accelerates as it keeps moving. In this case, the environment of the parcel is said to be unstable.

In the science of weather, when we speak about the stability/instability of a local environment, we are most of the time talking about the vertical stability/instability. With this in mind, consider our search of a developing thunderstorm. If some natural force causes parcels of air to start rising, and then the parcels keep rising on their own, or keep rising thanks to the initial force plus being able to rise on their own due to their own buoyancy, then eventually, if these parcels contain sufficient moisture in the form of water vapor, they will cool to their dewpoint, which starts forming cumulus-type clouds. Cumuloform clouds are clouds of significant vertical extent.

In an unstable environment, the air keeps accelerating upwards until it reaches a stable region aloft (this could be as high as the stratosphere if the whole troposphere were unstable!). Once the parcels have risen into a stable region aloft, they decelerate and eventually stop rising, which visibly would be seen as the tops of the convective clouds.

In other words, these blobs of air, which in strong convection are forming invisible currents of rising air, are buoyant when they are accelerating upwards on their own. To be buoyant, the parcels must be warmer than the environment through which they are rising. We refer to these rising parcels as being POSITIVELY BUOYANT. When the parcels then become colder than the environment (because they expend energy in the form of using up heat as they rise), they then sink, and are called NEGATIVELY BUOYANT.

The figure below illustrates stability and instability of air parcels.

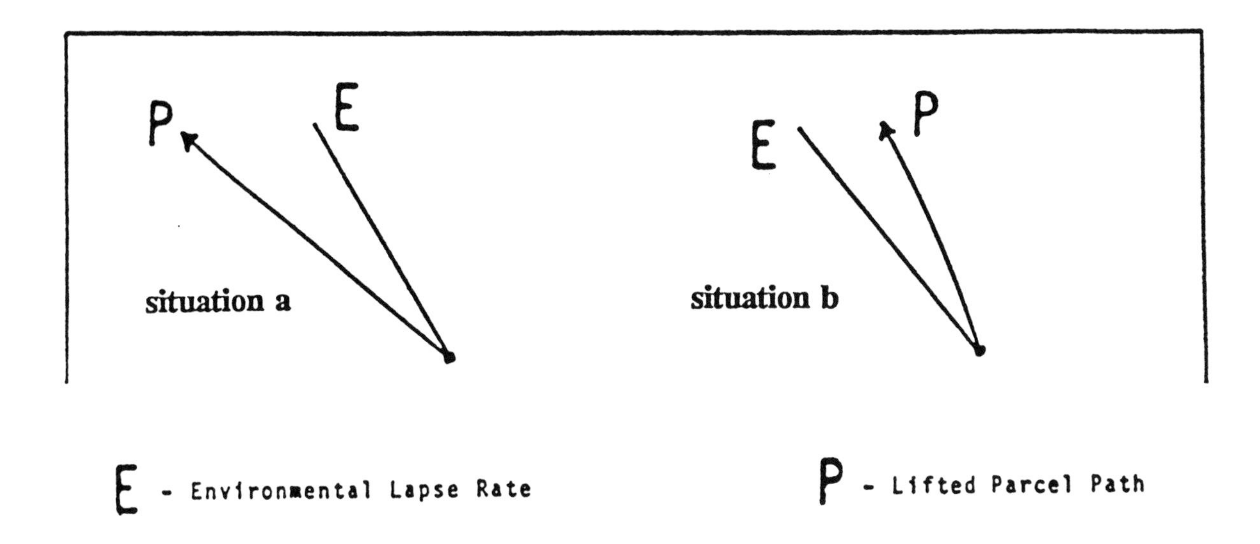

Figure 3-1. Stability vs. instability. E represents the environmental temperature lapse rate; i.e., E shows the actual temperature of the air (a falling temperature) as we ascend. P shows how an air parcel that is rising or is being lifted, cools as it ascends. The parcel cools at a specific rate, as explained later. In situation a, the parcel would be cooler than the environment, so it would sink and not form thunderstorms. The environment in situation a is said to be stable. In situation b, the parcel, although cooling as it ascends, remains warmer than the environment so it keeps rising. With moisture and cool enough air aloft, showers and/or thunderstorms may form.

Meteorologists routinely examine plotted vertical "soundings" taken by weather balloon instrument packages known as radiosondes, to predict the likelihood of thunderstorms. These plots of temperature, dewpoint and wind and various altitudes allow the forecaster to determine if the air is stable and unstable, and to what magnitude. The more unstable the air, the more likely thunderstorms are. Very unstable air often results in severe thunderstorms that produce downpours, frequent lightning, strong to damaging winds, hail and sometimes tornadoes and downbursts or microbursts of descending air at high speeds.

In a stable atmosphere, the temperature will not drop off as fast when we rise up through is, as it does in an unstable atmosphere. Since rising parcels give off energy in the form of heat when they rise, they cool at a certain rate as they are ascending. In a stable environment, these cooling parcels will be colder than the environment and cease to rise or even sink. In an unstable environment, these rising parcels, although cooling as in the stable condition, keep rising because the environmental temperature is colder than the temperature of the parcels. That it, the rate of decrease of the environmental temperature is greater than in the stable case. The result is that the parcels, even though they are cooling as they rise, remain positively buoyant; that is, they are warmer than the environment they are passing through, and therefore they continue to rise.

In stable conditions, we sometimes find cooler air hugging the lowest layers of air near the surface, and somewhat milder air immediately above it. This can also happen on clear, nearly calm nights when the air cools to its dewpoint and fog forms. In stable conditions, therefore, we find fog, but can also have prolonged periods of haze and pollution build-up since the milder air over relatively cool air traps the surface-based polluted air in place. Thus, prolonged stable conditions can cause a gradual build-up of pollutants in the air that we breathe, and we wait for a weather system such as a cold front or strong low pressure system to move in to stir up (mix) the air, and disperse the pollution.

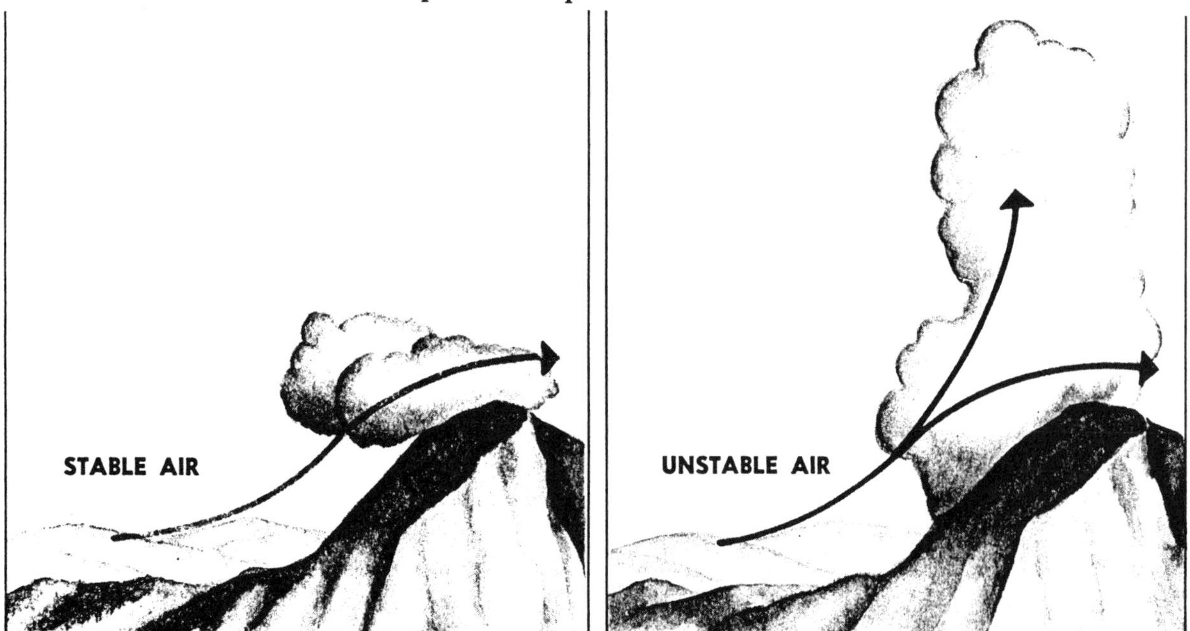

Figure 3-2. The kind of clouds we get in stable and unstable conditions. The figure shows moist air ascending a mountain. When the air is stable, the cloud type formed is called stratiform, such as with stratocumulus clouds, and has little vertical development. The clouds are more "stratified" than vertical. When the air is unstable, the cloud type formed is called cumuloform, such as with cumulus or cumulonimbus clouds, which have significant vertical development. (source: NOAA)

Stratified clouds yield steady precipitation, such as rain or snow, and cumuloform clouds yield showery precipitation, such as rain showers, snow showers and thunderstorms.

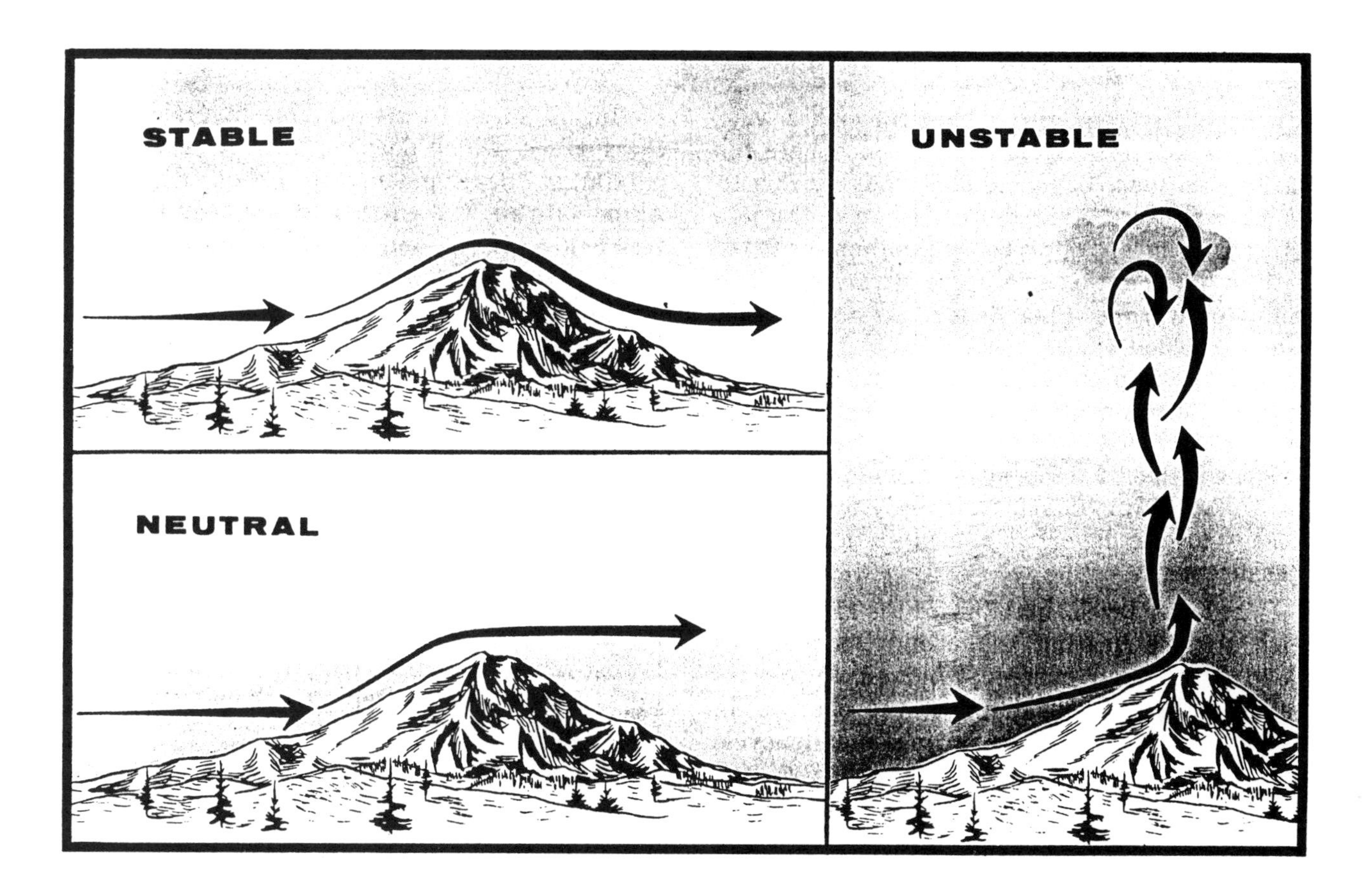

Figure 3-3. The stability of the air in the local area: stable, neutral or unstable. The arrows indicate the air flow. In these examples, air is forced to rise as it moves up over higher terrain. Sinking air, found in many stable situations, is called SUBSIDENCE. On flatter terrain, a common cause of air lift is heating of the ground by the sun, which heats the ground which then heats the air parcels above it. Action by fronts moving in is another source of lift. Even over the mountains, as shown in this figure, solar heating and frontal action are contributing agents for lift. Any weather features or combination of weather and terrain features that cause air parcels in the low levels of the troposphere to converge, will force the air to rise, since through convergence the air parcels must go somewhere after piling up, and at the surface they can go only upwards. (source: DOA)

This discussion on stability and instability is continued in greater detail in the next chapter.

Chapter 4. THE BIRTH OF A THUNDERSTORM

Figure 4-1 (previous page). A thunderstorm with vertical depth of some nine miles (some 14½ kilometers). (source: NOAA)

On a hot and humid summer afternoon you may be able to observe's Nature's artistry as you watch a puffy cumulus cloud grow upwards as if to be rising from a cauldron, into a cauliflower-shaped cumulonimbus cloud which is the thunderstorm. Then comes the dynamic display of lightning and thunder, with heavy sheets of wind-swept rain and sometimes stones of ice. On rare occasions, a funnel of intensely swirling air descends from typically the southwest part (in the Northern Hemisphere) of this thunderstorm, capable of performing massive damage and destruction. In fact, the rains are capable of producing destruction themselves, in the form of rushing waters in a flash flood, and the hail can pummel crops into the ground and smash windshields on automotive vehicles. The lightning is not only dangerous to people and animals, but can start fires that can destroy buildings and stands of forest. The lightning, the flash floods, the hail and the tornadoes are all life-threatening. Thus, books and videos on thunderstorms and their products will continue to be written, with updated information and knowledge.

When we consider some of the observations and reports of the effects of thunderstorms, it is understandable why we want to learn more about them.

For example, lightning has been known to hit telephone wires in the same community where two teenage girls were on the phone, come through the fires and kill both of them.

In a dry summer 1988 in northwest Wyoming, lightning started forest fires that burned approximately one million acres of that 2.2 million-acre National Park. The fires destroyed much of the older growth and underbrush, allowing new seeds to germinate with new evergreen trees and other vegetation renewing on the landscape. The smoke form the fires was carried by the winds as far away as Kansas City, Missouri, causing the sky on one late summer's day to be brown and obscuring most of the sun, even though there were no clouds.

A special form of lightning in the shape of a small ball, about the size of a basketball, has come into people's houses, then proceeding to chase a person or pet around the house before fizzling out. This "ball lightning" appears to be attracted to humans and larger animals when they are moving.

Flash floods have deluged local areas with as much as and even more than fifteen inches (over 380 millimeters) of rain within a few hours. For example, in the summer of 1976 such a flash flood --caused by some 15 inches of rain within a few hours-- killed scores of campers in the Big Thompson Canyon of Colorado on the night of July 31st, as a wall of cascading water swept through the canyon, sweeping people, campsites and vehicles downstream in a frenzy of destruction.

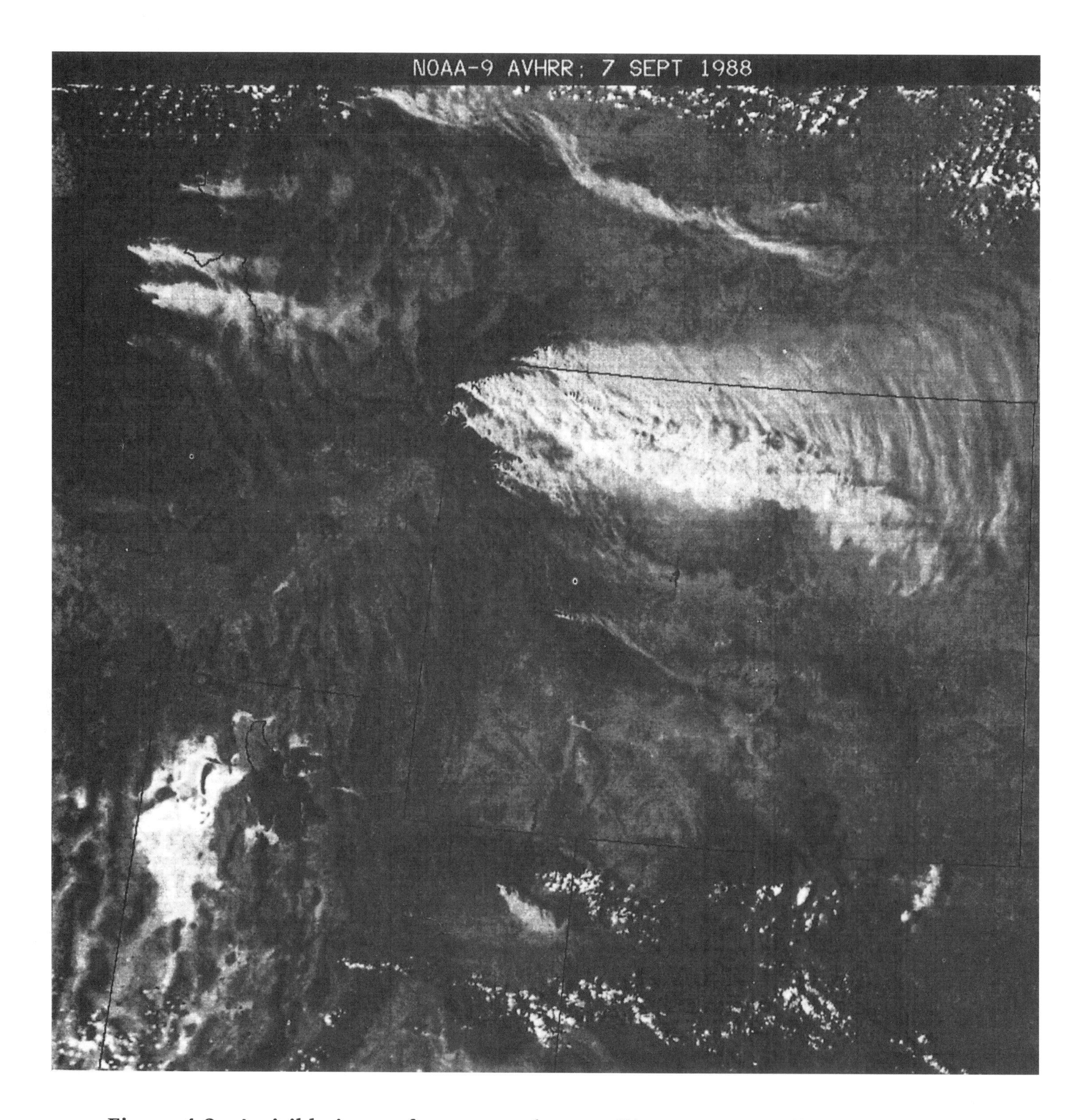

Figure 4-2. A visible image from a weather satellite captures Yellowstone National Park in northwest Wyoming burning due to lightning-induced forest fires. The brown smoke from Yellowstone and from another fire in western Montana are being blown towards the east-southeast by the winds. In the lower left is the Great Salt Lake in Utah and to the left (west) of it are the Bonneville Salt Flats. (source: NOAA)

Hail in summer thunderstorms has piled up several feet high. Hailstones themselves have reached, in the extreme, sizes greater than five inches across. Consider being caught outside when hailstones as big as softballs start pelting you, falling at high speeds from the sky. And think about what would happen to your car and windshields during such a hail episode.

Figure 4-3. A very large pinnacled hailstone that is sliced in two to show the layers of ice that comprise it. (source: NCAR)

Figure 4-4. This is a picture of a moderate-sized tornado, hundreds of yards in diameter. Even though it is not one-quarter to one-half of a mile wide, it is still potentially lethal as winds in even a smaller tornado can easily exceed 200 miles per hour. Winds do exceed 300 mph (do exceed 480 kilometers per hour) in the most extreme twisters.

Typically, a thunderstorm that contains large hail is more likely to produce a tornado than a non-hail thunderstorm. When hail of about one inch (2½ centimeters) in diameter or larger is occurring, watch the area to its southwest, for that is where, if a tornado is to occur, it will most likely be found in the thunderstorm. (source: NOAA)

a. necessary conditions for convection

A thunderstorm is a storm containing thunder and lightning. It forms when air rises or is forced to rise. These rising air parcels carry their warmth with them, although they cool as they rise: this process is known as convection.

Introduction:

We define a thunderstorm as any storm in which thunder is heard. Other storms, e.g., snowstorms and hurricanes, can also have thunderstorms imbedded in them; in fact, since hurricanes are convective, they originate as areas of thunderstorms and typically have thunder in them.

Thunderstorms used to be called "electrical storms", and perhaps more appropriately should be called lightning storms rather than thunderstorms. The term "thundershower" is obsolete; it used to refer to a relatively mild thunderstorm with little rain and wind; now, however, all electrical storms are referred to as "thunderstorms", since even a weak one produces lightning, and all lightning is potentially lethal.

All thunderstorms are accompanied by lightning. In fact, if it were not for the lightning, there would be no thunder. Thunder is caused by the rapid expansion of air once it is heated by lightning.

If you have seen a Van de Graaff generator (a device used to study static electricity), the crackling sound that you hear is really thunder, which is caused by the sparks (lightning) heating the air and the air subsequently expanding rapidly.

Thunderstorms are often accompanied by heavy rainfall and gusty winds. Sometimes they produce hail and occasionally they generate tornadoes.

A "severe thunderstorm" is a thunderstorm which produces either winds of 58 mph or more, or hail of 3/4 inch diameter or more, or both. Thus, severe thunderstorms are damaging.

Most of us think of thunderstorms as local storms that do not last too long, forming from large, swelling cumulonimbus clouds that look like giant cauliflowers. The diameter of most individual thunderstorms is about 5 to 6 miles. The vertical extent of these storms can grow to be over 50,000 feet high (about 10 miles), and even higher. In general, the taller the cloud, the more severe the thunderstorm is likely to be. During the daytime, the sky becomes noticeably darker as a thunderstorm moves in. This is because the storm's clouds are so thick (up to 10 miles or even more), that up to about 92% of the sunshine hitting its top is reflected back to space. Thus, not much of the sun's rays is making it through the thunderstorm, although there is also some scattered and reflected ambient light making it to the earth from the storm. The result, however, is that the sky turns much darker as a well-developed thunderstorm moves in.

In the warmer half of the year in mid-latitudes, showers become thunderstorms when the cloud tops exceed about 25,000 feet above the surface. The height of the thunderstorm cloud tops is impressive: they can grow to be over 40,000 feet high, with many of the most severe thunderstorm tops exceeding 55,000 feet. Tops are higher in the tropics than in the mid-latitudes, and higher in the mid-latitudes than in the polar regions; this is because the tropopause is highest in tropics, and becomes progressively lower as we move poleward. The rising air parcels no longer accelerate upward when they move into a stable region, such as the stratosphere which lies above the tropopause. When these rising parcels become cooler than the environment they are passing through, they slow down their ascent and eventually sink. Thus, thunderstorm cloud tops can grow only to a certain height. Moreover, eventually the moisture would be used up, which would mark the end of the cloud growth.

When a thunderstorm occurs during the daytime, we often do not see all the lightning until the cloud that produced it is very close to being overhead. This is because it is light outside. At night, however, we see the lightning before we hear its thunder. If someone shines a flashlight in your window on a bright sunny day, then you probably would not notice it unless you were looking out of the window; if this happens at night, then you would certainly notice it even if you were not facing the window. If we see lightning at night but do not hear its thunder, then we are observing what is called "distant lightning". Some people refer to this as "heat lightning", but this is a misnomer: there is no such thing as "heat lightning", or lightning produced by heat with no clouds; it is lightning observed at night from a distant thunderstorm. Typically, you do not hear thunder from a thunderstorm that is more than about 18 miles (29 kilometers) away. You may see only lightning if the storm is beyond about 18 miles or passes your area at this distance. As the storm gets closer, you may see the lightning and hear the thunder.

What comes first, the lightning or the thunder?: The lightning comes first, followed almost instantaneously by the thunder as the heated air rapidly expands in all directions from the lightning.

The reason we see the lightning before hearing the thunder is that the light from the lightning moves at the speed of light, about 186,000 miles per second, so that it is essentially instantaneous for us to see the lightning as it happens; however, the thunder moves at the speed of sound, which is about one-fifth of a mile per second. Therefore, it takes the thunder five seconds to travel one mile.

DEMONSTRATION: DETERMINING HOW FAR AWAY AN INDIVIDUAL THUNDERSTORM IS FROM YOU

The next time a thunderstorm is forming and approaching you, do the following:

1. Count the number of seconds between when you see the lightning and hear the thunder.

(continued)

2. Divide the number of seconds by five. Each multiple of five means the storm is one mile away.

3. Keep doing this as the storm approaches and then leaves. When the interval of time between when you see the lightning and hear the thunder decreases, the storm is getting closer; when this time interval increases, the storm is moving away.

Thus, if it takes 15 seconds between seeing the lightning and hearing the thunder, then the storm is 3 miles away. This works for individual thunderstorms. When there are clusters of thunderstorms, you may be seeing lightning and hearing thunder from more than one storm, so that you may not be able to do this demonstration. Moreover, storms sometimes form right over your location, and storms grow and die. When you see lightning and hear thunder about simultaneously from an individual thunderstorm cell, then the storm is directly overhead.

THE NECESSARY CONDITIONS FOR THUNDERSTORMS:

All four conditions must occur simultaneously:

1. LIFT
2. INSTABILITY
3. MOISTURE
4. LACK OF A MID-LEVEL CAP AROUND 10,000 FEET ABOVE THE SURFACE (it cannot be too warm aloft)

Let us look at each of these four conditions.

b. lifting mechanisms

To form the cumulus clouds that will grow into towering cumulus and ultimately the cumulonimbus clouds that give us thunderstorms, the air must rise or be forced to rise. This is because clouds form when moist air rises and cools to its dewpoint. As the air rises, it carries its moisture content with it except for some mixing in (entrainment) of environmental air. This means that the parcel's dewpoint is almost the same when the saturation elevation is reached as where the parcel started its ascent. Actually, there is some slight drop in dewpoint due to entrainment and also lower air pressure on the vapor causing some loss outward of moisture from the parcel.

Air parcels expend energy in the form of heat as they rise, so they cool at a certain rate. When these parcels are not yet saturated, they cool as what is called the DRY ADIABATIC LAPSE RATE. The word "adiabatic" means that no heat is being added to or taken away from the parcel; thus, the parcel is expending its own heat as it rises. This rate is about 5½ Fahrenheit degrees per 1000 feet of ascent (about 9.8 Celsius degrees per 1000 meters of ascent). However, once the air parcel has cooled to its dewpoint (is saturated; i.e., the relative humidity is now 100%), it continues to cool as it rises, but now at less of a cooling rate as before. This cooling rate when saturated is known as the MOIST ADIABATIC LAPSE RATE, which varies with height, but is about 60% of the dry adiabatic lapse rate. The moist adiabatic lapse rate in the low and middle troposphere is about 3F° per 1000 feet of ascent (about 6C° per 1000 meters of ascent).

The reason the air parcels cool at a slower rate once saturated is that since they are already saturated, any additional cooling releases the heat of condensation (the energy used by the parcel to convert the water vapor into liquid). By adding some heat to the parcel, the parcel does not then cool at its original rate.

This convective lifting is not the same as the synoptic lifting for widespread rainstorms and snowstorms caused by winter-type low pressure systems. The gradual lifting in these low pressure systems is over hundreds and hundreds of miles and is on the order of a few centimeters per second (less than one-fifth of a mile per hour). Compare this to the convective lifting in localized showers and thunderstorms, in which the rising air currents (updrafts) are on the order of 30 to 50 miles per hour, and are much greater in severe and hail-producing storms. Indeed, to generate softball-sized hailstones, the updrafts in a thunderstorm need to be about 150 mph!

The lifting mechanisms for thunderstorm development are characterized as dynamic lifting mechanisms and mechanical lifting mechanisms. The following are examples of some common types of lift for convection.

<u>Some dynamic lifting mechanisms for convection</u>:

- Heating of the day. The sun heats the ground, which in turn heats the air above it, which rises as parcels of warm air.

- An advancing cold front. The leading edge of the colder air at the surface scoops up the warmer air just ahead of it. Sometimes the cold front slows down at the surface but some of the colder air advances aloft, say at about 5,000 to 10,000 feet up. This destabilizes the mid-levels of the troposphere, which enhances convective lift.

- On the cool air side of an advancing warm front, especially if the warm air is moist and unstable. Steady precipitation with imbedded showers and thunderstorms often occur ahead of and along the warm front.

- Intersection of low-level boundaries. Intersections of any two boundaries can create lift. Such boundaries include the outflow of air from the leading edge of a thunderstorm (or dying thunderstorm) intersecting the outflow of air from another thunderstorm. These outflows are termed "outflow boundaries". Another example is a sea-breeze front (or lake-breeze front) intersecting another boundary. On a peninsula, for example, two sea-breeze fronts from two different directions may bang into each other, resulting in cumuloform clouds forming there, which may grow into thunderstorms if the other necessary conditions for thunderstorms are present.

Some mechanical lifting mechanisms for convection:

•Upslope. This occurs when the low-level air is forced to rise up over gradually rising terrain. An example is an easterly flow of air (air moving westward) from central Kansas to eastern Colorado. If sufficient moisture is present in the air, it will cool to its dewpoint as it rises, and if the environment is unstable, the clouds will grow.

•Orographic lifting. Clouds may form when air is forced to ascend mountains. Upslope is one type of orographic lifting.

In each of these lifting episodes, the air that is rising or being lifted must have ample moisture to generate clouds once the air is cooled to its dewpoint temperature. Keep in mind that the dewpoint is in reality a measure of the amount of moisture in the air. The higher the dewpoint, the more grams of moisture per kilogram of air. Thus, many of the heaviest rainfalls in thunderstorms occur from storms (especially slow-moving ones) that form in air with dewpoints of 70 degrees F. or higher.

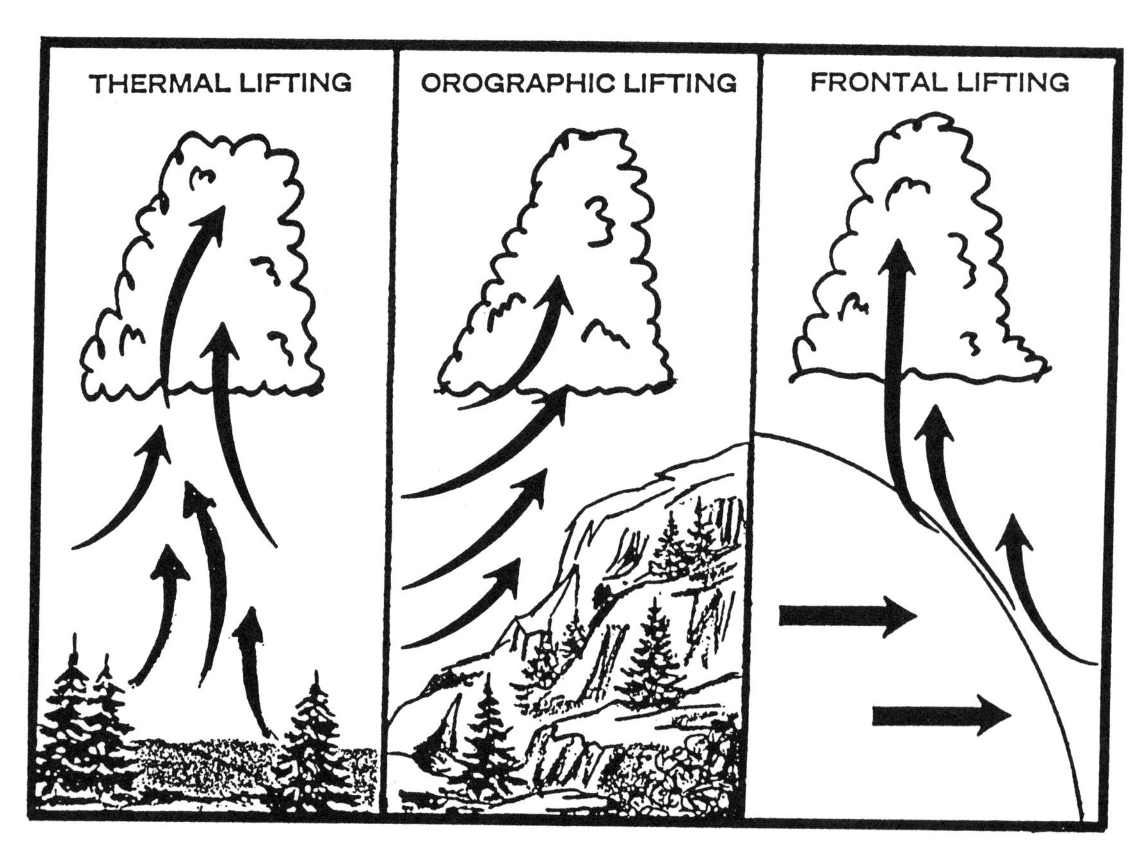

Figure 4-5. Examples of some of the lifting mechanisms that would produce thunderstorms if they occur in moist and unstable air, without a cap (too warm) aloft existing about 10,000 feet off the ground. (source: DOA)

Figure 4-6. As a cold front moves through, it scoops up the warmer air ahead of it. If the warmer air is also moist and unstable without a cap aloft, thunderstorms are likely just ahead of and along the front. Sometimes a cold front slows down as it tries to dislodge hot air, but some of the cooler air advances aloft at some 5,000 to 10,000 feet above the ground. This destabilizes the mid-troposphere and can lead to a line of thunderstorms some 100 miles ahead of the front (a pre-frontal squall-line), while along the cold front there is another line of thunderstorms. (source: DOA)

Figure 4-7. An advancing warm front first has the warmer air coming in aloft over the cool or cold air, since the warmer air is lighter than the cool or cold air. Where the warm air is at the surface is where the warm front is located on the surface weather map. If the warm air is also moist and unstable, imbedded showers and thunderstorms are likely within the general precipitation area on the cold air side of the warm front. (source: DOA)

c. specifics on instability

Chapter 3 introduced the subject of stability and instability in the atmosphere. By way of a quick review, recall that in a fluid, such as in the atmosphere or in the ocean, if a small parcel of that fluid is given a push sideways or vertically or somewhere inbetween horizontal and vertical, and once the pushing agent ceases, the parcel then returns to its location of origin, then that environment is said to be stable; however, if that parcel continues moving away from its origin, then that environment is said to be unstable.

For convection to proceed to the development of a thunderstorm, the local environment must be unstable in the vertical. Therefore, as the temperature of the environment falls with height, and even though the temperature of the rising parcel is dropping as it ascends, the comparison of the two temperatures must be that the air parcel's temperature remains higher than that of the environment it is passing through. Then the air parcel is said to be **POSITIVELY BUOYANT** and it keeps rising, even accelerating. The acceleration and then rise stops when the parcel becomes **NEGATIVELY BUOYANT**, which means that it is cooler than the environment. We are speaking about parcels of from about a cubic foot to a cubic meter in volume.

If one of the four conditions for thunderstorms is missing, there will not be any thunderstorm formation. The crucial fact that we have learned through thunderstorm and severe weather research is that we MUST have all four conditions for thunderstorm development coexisting in the same area: a lifting mechanism, instability, sufficient moisture and lack of a mid-level cap.

•Suppose there is no lifting mechanism at work. Then, even with great instability, copious moisture, and no cap aloft (no air too warm in the mid-troposphere), we will not even get any cumulus clouds.

•Suppose we have a lifting mechanism or a combination of lifting mechanisms at work, adequate moisture to form clouds, and no cap aloft, yet the environment is stable. Then if we have clouds, they would be the stratiform type, such as stratus or stratocumulus, but we will not have convection.

•Suppose we have lift, an unstable environment, and no cap aloft, but the moisture supply is very limited, such as in early summer over the southern Arizona and southeast California desert. Then the rising air currents comprised of the rising air parcels will occur, but little if any cloudiness will form. Sometimes there is moisture in low or mid levels of the troposphere only, rather than through a deep layer. There needs to be sufficient moisture to permit cloud formation through a deep vertical rise. When adequate water vapor in the air exists in mid-levels (some

6500 to 20,000 feet off the ground), then the bases of thunderstorms may start that high up, rather than some 2 thousand to 5 thousand feet off the ground, which height is common for many thunderstorm cloud bases initially (the bases then lower as the storm intensifies). When the bases do form higher, such a thunderstorm is referred to as a **MID-BASED THUNDERSTORM.** In drier parts of the country, some or all of the precipitation from such storms may evaporate before reaching the ground. Moreover, the downdrafts associated with the falling precipitation may accelerate downward from the cloud, leading to strong and sometimes damaging downbursts of air. Instability can also lead to parcels of air accelerating downward towards the ground when the initially descending parcels are cooler than the environment and, even though they are warming as they descend, as long as they remain cooler than the environment through which they are falling, these air parcels are then forming a current of air that keeps descending faster and faster, reaching the ground as high winds coming from above.

When air parcels sink, or when an entire layer of air sinks, this sinking is called **SUBSIDENCE.**

●Suppose we have lift, instability, moisture but a cap (sometimes called a "lid" aloft in mid-levels of the troposphere, around about 10,000 to 12,000 feet, for example. Then the convective (cumuloform) clouds will form and grow since the parcels are rising, but as soon as these parcels hit the relatively warm layer aloft, they will be cooler than the air in that layer and will decelerate and eventually stop rising and then sink, since they will be cooler than the environment. This cap areas is therefore an area of negative buoyancy. What we may see visibly are developing cumulus clouds that may start growing into towering cumulus clouds, but then the tops of these clouds spread out horizontally, forming little anvils, as the rising moist air parcels stop rising. In the middle latitudes during the warmer part of the year, showers will become thunderstorms (the first lightning and thunder occur) when the cloud tops reach about 25,000 feet high, above the land or body of water surface. The cap is also called a **CAP INVERSION** since it is a layer through which the temperature is usually rising with increasing height. The cap inversion may be from one to two thousand feet deep to several thousand feet deep. A thin cap, say one under 2000 feet thick, can be broken by rapidly-rising strong updrafts, but a deeper one, especially when the temperature increase is several Celsius degrees, will more likely inhibit convective air currents from breaking through it. A thin cap may also be conducive to severe thunderstorm development, and more likely for rapid thunderstorm-genesis. This is because the energy of the updrafts keeps trying to break the cap and does not, but suddenly, and especially when the air below the cap becomes more unstable, the updrafts burst through the thin cap, resulting in the explosive development of thunderstorms. The "lid" is broken. In mid-latitudes over non-mountainous terrain during the warm season, a 700 millibar (hectoPascal) (about 12,000 feet above sea-level) temperature of 12 degrees Celsius or higher is usually sufficient to inhibit thunderstorm development. A 14°C or above temperature virtually assures thunderstorm inhibition.

Thus, we know for sure that we must have lift, instability, moisture and no cap aloft

An important point about unstable air is that we are assuming that there is essentially no heat added to or taken away from these moving parcels by the environment. Then the process is called **ADIABATIC**, which means no heat transfer in this process.

Two necessary points to keep in mind when studying this topic of stability/instability in the atmosphere are: pressure in the atmosphere decrease with height, and the temperature of small parcels of air decreases as the air expands and rises, as long as the process is adiabatic (no heat is given to or taken from the parcels). The converse of these two points is also true.

The figure below illustrates air cooling as it rises, and air warming as it sinks.

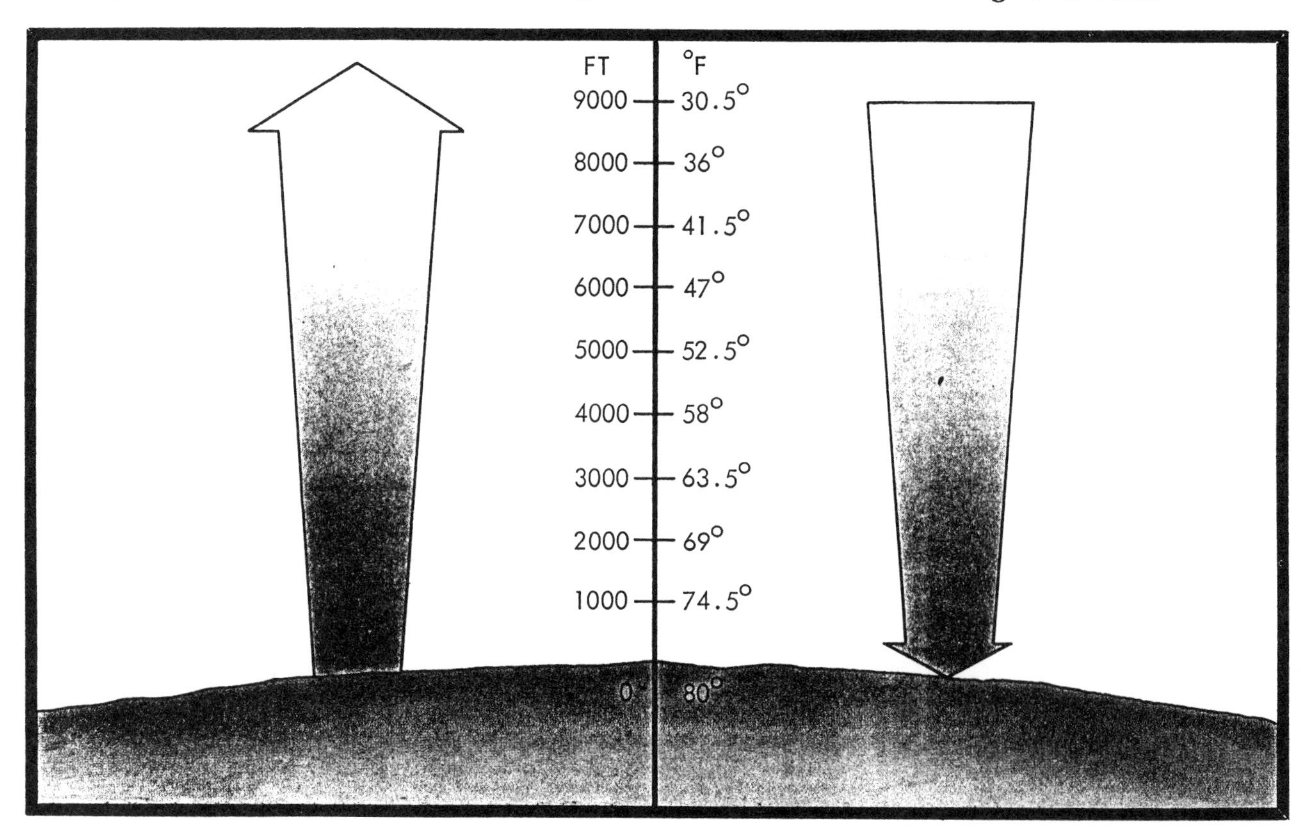

Figure 4-8. Rising air expands and cools since it uses up energy in the form of heat as it rises. Sinking air is compressed and warmed. When no heat is added to these parcels from the environment, then the rising and sinking motions are called adiabatic. (source: DOA)

Let us look at the average temperature profile of earth's atmosphere. Notice from the figure below that the temperature in general, the temperature decrease with height in the areas called the troposphere and the mesosphere, and increase with increasing elevation in the stratosphere and thermosphere. Then higher up, the temperature cools off again as the gas particles that comprise our atmosphere thin out and we then reach outer space, which starts at from 600 to 100 miles out from the earth's surface. It is easiest for clouds to form in the troposphere since the best combination of transient areas of instability and abundant water vapor in the air are found in the troposphere.

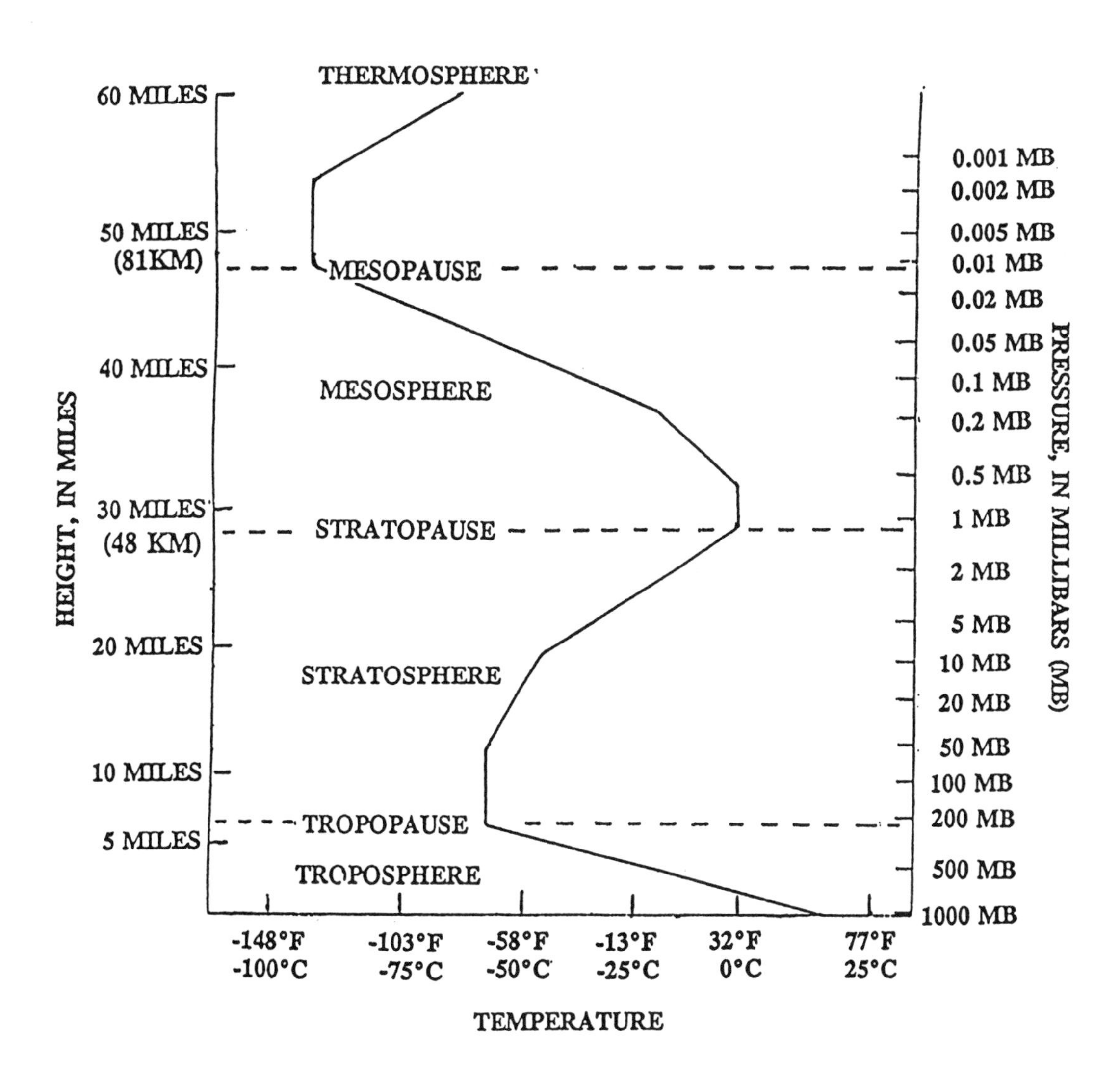

Figure 4-9. The vertical thermal profile of earth's atmosphere, not drawn to scale. Half the contents of the atmosphere are in the lower 18,000 feet above the surface, and the other half of the atmosphere is above 18,000 feet up to 600 to 1000 miles up, which is the top and end of our atmosphere.

Now let us look more closely at some aspects of stability and instability in the atmosphere.

Stable air masses and temperature inversions:
Temperature inversions occur when a layer of warmer air is on top of a layer of relatively cooler air. This is called a stable to very stable condition, because the cold air, which is heavier than the warm air, tends to remain close to the surface, and the lighter warmer air tends to remain above. As a result, the pollutants are trapped below in the colder air hugging the earth's surface, and we have a pollution build-up as more automotive emissions and the products of burning are released into the lowest layer of the atmosphere. Temperature inversions are most likely on clear, calm or nearly calm nights, when heat from the surface is lost due to radiational cooling.

When the air is stable, with little if any upward vertical motion, there may be a pollution build-up. Fog occurs under stable conditions. A prolonged period of fog can also lead to fog plus pollutants hanging over the area.

When air is unstable, we find rising air currents. This condition may lead to showers and thunderstorms. Locally, where this is occurring, pollutants are unlikely to build up.

Temperature inversions are most likely in valleys. During a clear, calm night, the radiational cooling and the sinking of the denser cool air towards the valley bottom results in the temperature near the ground being colder than the temperature several hundred feet above.

Even in a local countryside area, you may walk late on a summer's evening down a hill into a low area and feel on your skin a sudden drop in temperature from what it was at the higher elevation.

In Donora, Pennsylvania in October 1948, several days of stable air and nearly calm winds led to a few days of persistent dense fog. Fumes from a steel mill, a smelter and other polluting sources "hung" in the fog and was breathed in by people there. Twenty-two people died from this extreme case of air pollution, and about 7000 became ill. Most of the victims were people with cardiac and respiratory problems, such as people who smoked cigarettes. This was one of the worst air pollution disasters in the United States.

The figure on the next page shows graphical plots of the rate of change of temperature with increasing height, which is called the temperature lapse rate. This is for the local vertical environment. As air parcels ascend, they cool at the dry adiabatic lapse rate if they are not saturated with all the moisture they can hold (at which point they would have 100% relative humidity, which means they are holding 100% of the water vapor that they can hold at that temperature and pressure), or, if they are saturated, they cool at the moist-adiabatic lapse rate. Thus, we have the lapse rate of the parcel and the actual environment air temperature lapse rate.

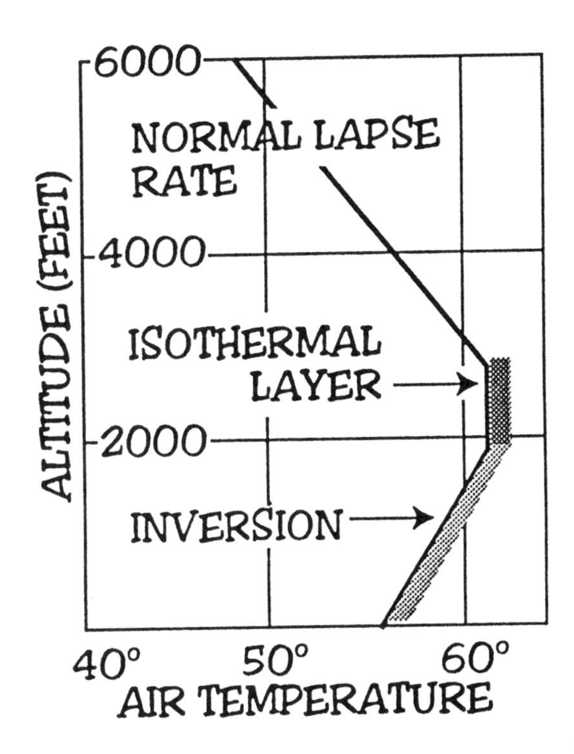

Figure 4-10. A figure showing temperature lapse rates. A lapse rate is the change of temperature with height. Normally, the air temperature gets colder as we rise in elevation until we reach the stratosphere. In this example, note that the temperature gets warmer from the surface to 2000 feet up. This is called a temperature inversion. "Isothermal" means the temperature remains the same with height. (source: USN)

Figure 4-11. Smoke that is released into an inversion layer rises until its temperature equals that of the surrounding air, at which point the smoke flattens out and spreads horizontally. Eventually, much of it may also work its way back downwards. (source: DOA)

Figure 4-12. A temperature inversion in a valley, showing the temperature at different elevations. The zone of warm nighttime temperatures near the top of the inversion is called "the thermal belt". (source: DOA)

Temperature inversions require clear or nearly clear skies so that the earth and air in contact with it in that region can radiate out into space the maximum amount of heat, to cool off the most possible. Clouds trap some of that outgoing radiation and reradiate it back to the surface; in other words, cloud cover keeps the area warmer than it otherwise would be at night. Moreover, if the winds are above about 6 miles per hour, then there will be some mixing of the nighttime air, and it will not cool as much. Thus, to have a nocturnal temperature inversion, we need very precise conditions: clear skies and near-calm or calm winds. If the air cools early enough to its dewpoint, dew will form and then fog. If the fog is very deep and becomes fairly widespread, then these stable conditions can remain for days, until a strong weather system comes in to disperse it. Under prolonged inversions, pollutants build up because they are trapped in that inversion air layer.

Figure 4-13. Visible indicators of a stable atmosphere. (source: DOA)

Figure 4-14. Visible indicators of an unstable atmosphere. (source: DOA)

When the local atmospheric environment is stable, then there is a resistance of the atmosphere to vertical motion. This explanation is based on the parcel method of analysis, relative to the vertical temperature and moisture (dewpoint) sounding thorugh the troposphere. (The rest of this section on "specifics on instability" comes from an excellent essay on the topic prepared by the U. S. Department of Agriculture:)

This method employs some assumptions: (1) The sounding applies to an atmosphere at rest; (2) a small parcel of air in the sampled atmosphere, if caused to rise, does not exchange mass or heat across its boundary; and (3) rise of the parcel does not set its environment in motion. We learned that lifting under these conditions is **adiabatic lifting.**

Three characteristics of the sounding then determine the stability of the atmospheric layer in which the parcel of air is embedded. These are: (1) The temperature lapse rate through the layer; (2) temperature of the parcel at its initial level; and (3) initial dew point of the parcel.

Adiabatically lifted air expands in the lower pressures encountered as it moves upward. This is a cooling process, and the rate of cooling with increase in altitude depends on whether or not the temperature reaches the dew point and consequent saturation. As long as the air remains unsaturated, it cools at the constant dry-adiabatic lapse rate of 5.5°F. per 1,000 feet of rise. Rising saturated air cools at a lesser rate, called the **moist-adiabatic** rate. This rate averages about 3°F. per 1,000 feet, but, as we will see later, it varies considerably.

STABILITY DETERMINATIONS

The degree of stability or instability of an atmospheric layer is determined by comparing its temperature lapse rate, as shown by a sounding, with the appropriate adiabatic rate. A temperature lapse rate less than the dry-adiabatic rate of 5.5°F. per 1,000 feet for an unsaturated parcel is considered **stable,** because vertical motion is damped. A lapse rate greater than dry-adiabatic favors vertical motion and is **unstable.** In the absence of saturation, an atmospheric layer is **neutrally stable** if its lapse rate is the same as the dry-adiabatic rate. Under this particular condition, any existing vertical motion is neither damped nor accelerated.

In the case of a saturated parcel, the same stability terms apply. In this case, however, the comparison of atmospheric lapse rate is made with the moist-adiabatic rate appropriate to the temperature encountered.

Layers of different lapse rates of temperature may occur in a single sounding, varying from superadiabatic (unstable), usually found over heated surfaces, to dry-adiabatic (neutral), and on through inversions of temperature (very stable). In a saturated layer with considerable convective motion, the lapse rate tends to become moist-adiabatic.

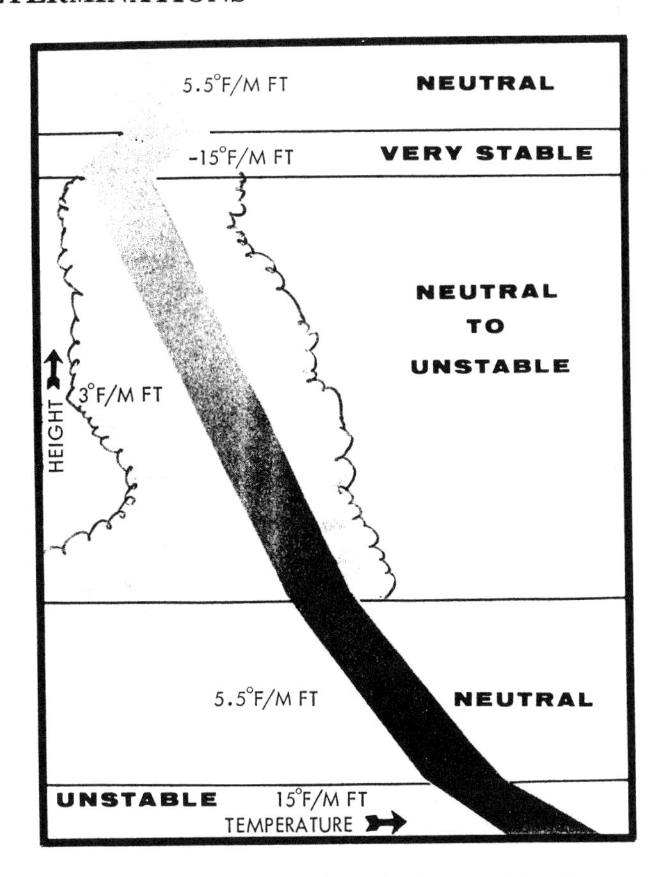

Atmospheric stability of any layer is determined by the way temperature varies through the layer and whether or not air in the layer is saturated.

Figure 4-15.

The adiabatic process is reversible. Just as air expands and cools when it is lifted, so is it equally compressed and warmed as it is lowered. Hence, adiabatic processes and stability determinations for either upward or downward moving air parcels make use of the appropriate dry— or moist-adiabatic lapse rates. The temperature structure of the atmosphere is always complex. As mentioned above, the moist-adiabatic lapse rate is variable—not constant as is the dry-adiabatic rate.

Adiabatic Chart

To facilitate making stability determinations, therefore, meteorologists analyzing upper-air observations use a thermodynamic diagram called an **adiabatic chart** as a convenient tool for making stability estimates. The basic portion of the chart is a set of gridlines of temperature and pressure (or height) on which the measured temperature and moisture structure of the atmosphere can be plotted. The moisture is plotted as dew-point temperature. Also printed on the chart is a set of dry-adiabatic and a set of moist-adiabatic lines. By referring to these adiabats, the lapse rates of the various layers or portions of the atmosphere can be compared to the dry-adiabatic rate and the moist-adiabatic rate. In later chapters we will consider other ways in which the adiabatic chart is used.

Stability determinations from soundings in the atmosphere are made to estimate the subsequent motion of an air parcel that has been raised or lowered by an external force. In a stable atmosphere, the parcel will return to its original position when the force is removed; in an unstable atmosphere, the parcel will accelerate in the direction of its forced motion; and in a neutrally stable atmosphere, it will remain at its new position.

Stability of Unsaturated Air

We can illustrate use of the adiabatic chart to indicate these processes by plotting four hypothetical soundings on appropriate segments of a chart. We will first consider only unsaturated air to which the constant dry-adiabatic lapse rate applies.

Assume for simplicity, that each of our four soundings has a lapse rate indicated dia-

Figures 4-16.

The ADIABATIC CHART

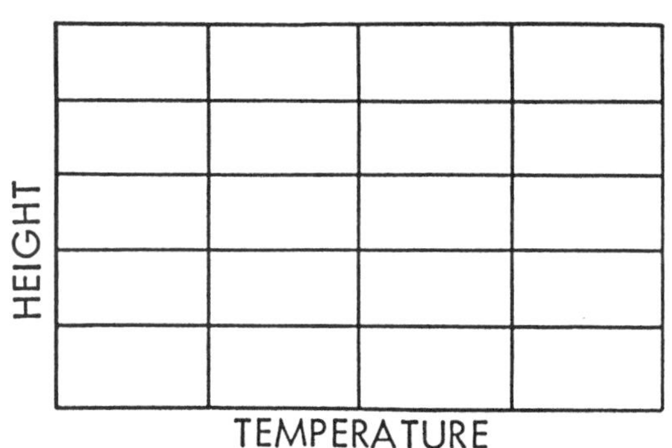

TO THE TEMPERATURE-HEIGHT DIAGRAM—

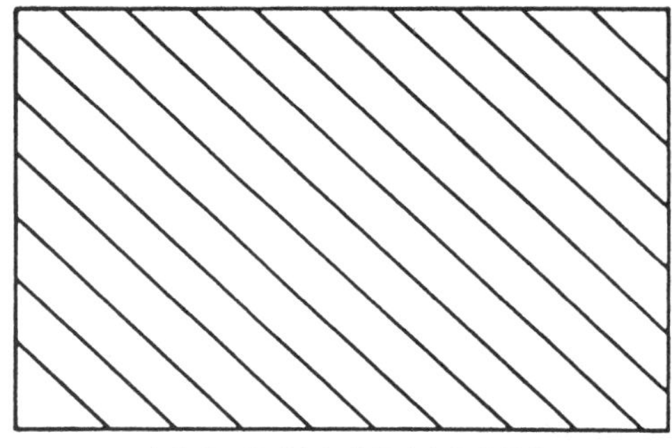

ADD DRY ADIABATS —

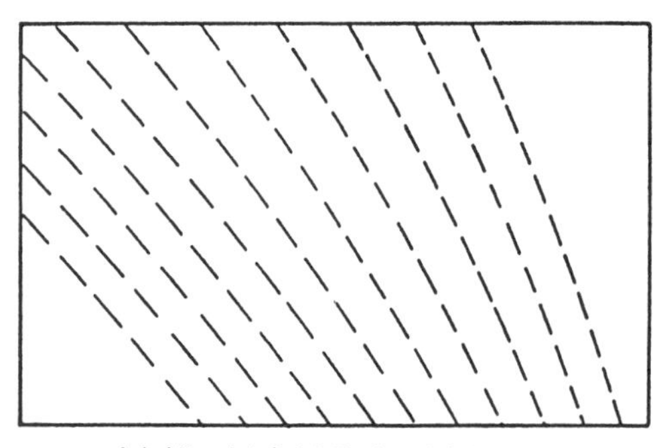

AND MOIST ADIABATS

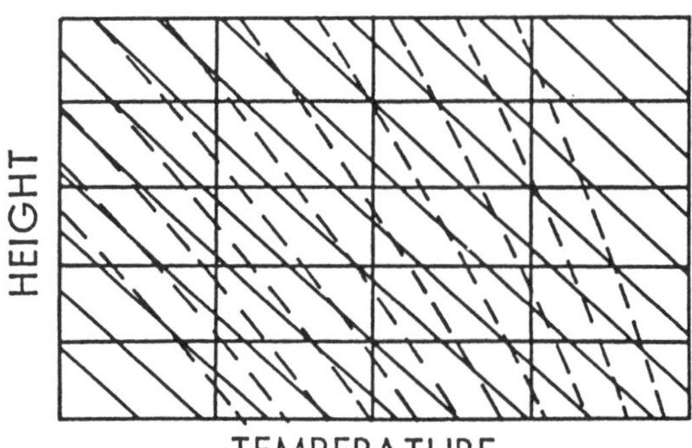

To determine stability, the meteorologist plots temperature and moisture soundings on an adiabatic chart and compares the lapse rates of various layers to the dry adiabats and moist adiabats.

grammatically by a solid black line. Note also in the accompanying illustration that each shows the temperature at 3,000 feet to be 50°F. For our purposes, let us select a parcel of air at this point and compare its temperature with that of its environment as the parcel is raised or lowered by external forces. If it remains unsaturated, the parcel will change in temperature at the dry-adiabatic rate indicated on the chart by red arrows.

The sounding plotted in (A) has a lapse rate of 3.5°F. per 1,000 feet. As the parcel is lifted and cools at its 5.5° rate, it thus becomes progressively colder and more dense than its environment. At 5,000 feet, for example, its temperature would be 39°F., but the temperature of the surrounding air would be 43°F. Gravity thus returns the parcel to its point of origin when the external force is removed. Moved downward, the parcel warms at the dry-adiabatic rate and becomes warmer than its environment. At 1,000 feet, for example, the parcel temperature would be 61°F., but the temperature of the environment would be only 57°F. Buoyancy forces the parcel back up to its original level. The damping action in either case indicates stability.

The parcel in (B) is initially in an inversion layer where the temperature increases at the rate of 3°F. per 1,000 feet of altitude. If the parcel is lifted, say 1,000 feet, its temperature will decrease 5.5°F., while the temperature of the surrounding air will be 3°F. higher. The parcel will then be 8.5°F. colder and will return to its original level as soon as the lifting force is removed. Similarly, a lowered parcel will become warmer than the surrounding air and will also return to its original level. Thus, inversions at any altitude are very stable.

Next, let us consider (C) where the parcel is embedded in a layer that has a measured lapse rate of 5.5°F. per 1,000 feet, the same as

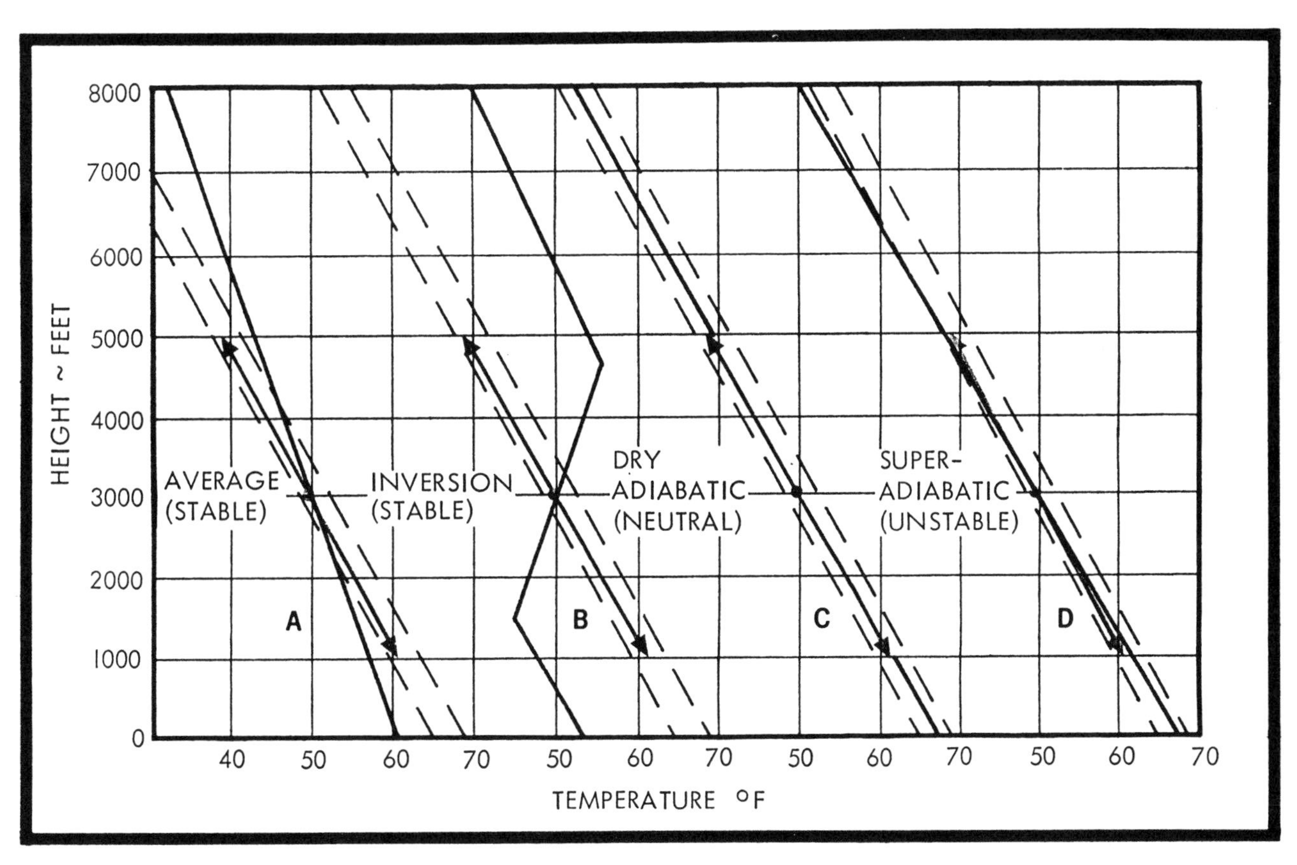

Figure 4-17. In unsaturated air, the stability can be determined by comparing the measured lapse rate (solid black lines) to the dry-adiabatic lapse rate (dashed black lines). The reaction of a parcel to lifting or lowering may be examined by comparing its temperature (red arrows for parcel initially at 3,000 feet and 50°F.) to the temperature of its environment.

the dry-adiabatic rate. If moved upward or downward in this layer, the parcel will change in temperature at the same rate as that of its environment and, therefore, will always be in temperature equilibrium with the surrounding air. The parcel will come to rest at its new level when external forces are removed. Technically, such a layer is neutrally stable, but we will see, after we consider an unstable case, that a neutrally stable layer is a potentially serious condition in fire weather.

In the last example (D) in unsaturated air, the plotted temperature lapse rate is 6°F. per 1,000 feet, which is greater than the dry-adiabatic rate. Again, if our parcel is lifted, it will cool at the dry-adiabatic rate or 0.5° less per 1,000 feet than its surroundings. At an altitude of 5,000 feet, for example, the temperature of the parcel would be 39°F., while that of its surroundings would be 38°F. Thus, the parcel is warmer and less dense than the surrounding air, and buoyancy will cause it to accelerate upward as long as it remains warmer than the surrounding air. Moved downward, the parcel would similarly cool more rapidly than the surrounding air and accelerate downward. Hence, an atmospheric layer having a lapse rate greater than the dry-adiabatic rate is conducive to vertical motion and overturning, and represents an unstable condition.

Lapse rates greater than the dry-adiabatic rate, we learned in chapter 2, are called superadiabatic. But since they are unstable, the air tends to adjust itself through mixing and overturning to a more stable condition. Superadiabatic lapse rates are not ordinarily found in the atmosphere except near the surface of the earth on sunny days. When an unsaturated layer of air is mixed thoroughly, its lapse rate tends toward neutral stability.

The term "neutral" stability sounds rather passive, but we should be cautious when such a lapse rate is present. The temperature structure of the atmosphere is not static, but is continually changing. Any warming of the lower portion or cooling of the upper portion of a neutrally stable layer will cause the layer to become unstable, and it will then not only permit, but will assist, vertical motion. Such changes are easily brought about. Thus, we should consider the terms stable, neutral, and unstable in a relative, rather than an absolute, sense. A stable lapse rate that approaches the dry-adiabatic rate should be considered relatively unstable.

Warming of the lower layers during the daytime by contact with the earth's surface or by heat from a wildfire will make a neutral lapse rate become unstable. In an atmosphere with a dry-adiabatic lapse rate, hot gases rising from a fire will encounter little resistance, will travel upward with ease, and can develop a tall convection column. A neutrally stable atmosphere can be made unstable also by **advection**; that is, the horizontal movement of colder air into the area aloft or warmer air into the area near the surface. Once the lapse rate becomes unstable, vertical currents are easily initiated. Advection of warm air aloft or cold air near the surface has the reverse effect of making the atmosphere more stable.

So far we have considered adiabatic cooling and warming and the degree of stability of the atmosphere only with respect to air that is not saturated. Rising air, cooling at the dry-adiabatic lapse rate, may eventually reach the dew-point temperature. Further cooling results in the condensation of water vapor into clouds, a change of state process that liberates the latent heat contained in the vapor. This heat is added to the rising air, with the result that the temperature no longer decreases at the dry-adiabatic rate, but at a lesser rate which is called the moist-adiabatic rate. On the average, as mentioned earlier, this rate is around 3°F. per 1,000 feet, but it varies slightly with pressure and considerably with temperature. The variation of the rate due to temperature may range from about 2°F. per 1,000 feet at very warm temperatures to about 5°F. per 1,000 feet at very cold temperatures. In warmer air masses, more water vapor is available for condensation and therefore more heat is released, while in colder air masses, little water vapor is available.

Stability of Saturated Air

Let us now consider a situation in which an air parcel is lifted and cooled until it reaches saturation and condensation. For this, we need to know both the initial temperature of the parcel and its dew-point temperature. This stability analysis of a sounding makes use of both the dry-adiabatic and moist-adiabatic lines shown on the adiabatic chart. For this example, assume a sounding, plotted on the accompanying chart, showing a temperature lapse rate of 4.5°F. We will start with a parcel at sea level where the temperature is 80°F. and the dew point is 62°.

The 80°F. temperature and 62° dew point indicate that the parcel is initially unsaturated. As the parcel is lifted, it will cool at the dry-adiabatic rate until saturation occurs. The parcel dew-point temperature meanwhile decreases, as we learned in chapter 3, at the rate of 1°F. per 1,000 feet. If we draw a line on the adiabatic chart with a slope of −1°F. starting at the surface 62° dew point, we find that this line intersects the dry-adiabatic path of the parcel. The parcel temperature at this point is therefore at the dew point. The altitude of the point is thus at the **condensation level.**

In our example, condensation occurs at 4,000 feet above sea level at a temperature of 58°. The atmosphere is stable at this point because the parcel temperature is lower than that shown by the sounding for the surrounding air. If the parcel is forced to rise above the condensation level, however, it then cools at the moist-adiabatic rate, in this case about 2.5°F. per 1,000 feet. At this rate of change, the parcel temperature will reach the temperature of the surrounding air at 6,000 feet. The level at which the parcel becomes warmer than the surrounding air is called the **level of free convection.** Above this level, the parcel will become buoyant and accelerate upward, continuing to cool at the moist-adiabatic rate, and no longer requiring an external lifting force.

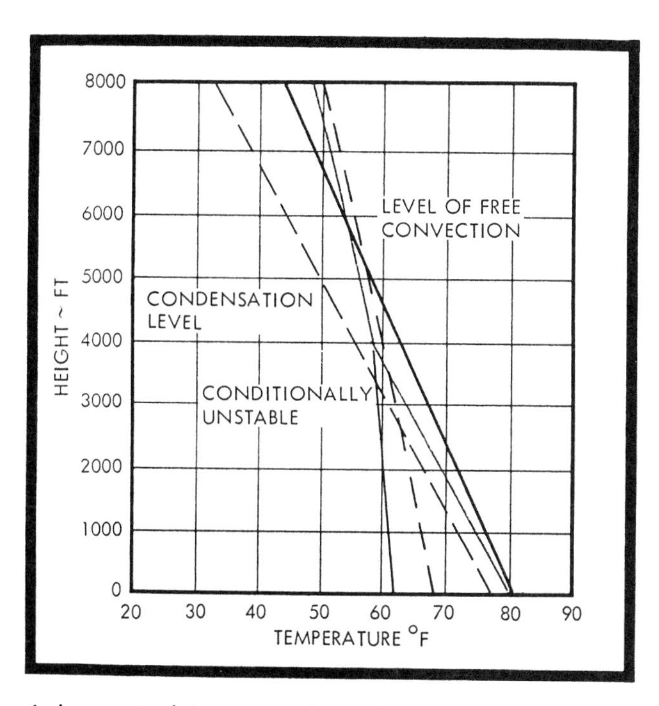

A lapse rate between the dry- and moist-adiabatic rates is conditionally unstable, because it would be unstable under saturated conditions but stable under unsaturated conditions. The temperature of a parcel raised from near the surface will follow the dry-adiabatic rate until saturation, then follow the moist-adiabatic rate. At the level where the parcel temperature exceeds the environment temperature, the parcel will begin free ascent.

Figure 4-18.

Conditional Instability

The atmosphere illustrated by the above example, which has a lapse rate lying between the dry and moist adiabats, is said to be **conditionally unstable.** It is stable with respect to a lifted air parcel as long as the parcel remains unsaturated, but it is unstable with respect to a lifted parcel that has become saturated. In our example, the measured lapse rate of the layer is 4.5°F. This layer is, therefore, stable with respect to a lifted parcel as long as the parcel temperature follows the dry-adiabatic rate. It is unstable with respect to a lifted saturated parcel, because the temperature of the saturated parcel would follow the lesser moist-adiabatic rate, in this case about 2.5°F. per 1,000 feet.

A saturated parcel in free convection loses additional moisture by condensation as it rises. This, plus the colder temperature aloft, causes the moist-adiabatic lapse rate to increase toward the dry-adiabatic rate. The rising parcel will thus eventually cool to the temperature of the surrounding air where the free convection will cease. This may be in the vicinity of the tropopause or at some lower level, depending on the temperature structure of the air aloft.

Reliance on the parcel method of analyzing atmospheric stability must be tempered with considerable judgment. It is true that from the plotted temperature lapse rates on the adiabatic chart one can read differences between temperatures of parcels and the surrounding air. These are based, however, on the initial assumptions upon which the method is founded. One of these, for example, is that there is no energy exchange between the parcel and the surrounding air. Vertical motion is, however, often accompanied by various degrees of mixing and attendant energy exchange, which makes this assumption only an approximation. The usual practice of plotting the significant turning points from sounding data and connecting them with straight lines also detracts from precision. These are additional reasons for considering stability in a relative sense rather than in absolute terms.

The temperature of the parcel and the environment, and the dew-point temperature of the parcel used in this example, are summarized below:

Altitude		Environment temperature	Parcel temperature		Dew-point temperature
Sea level		80	80	Dry-adiabatic lapse rate	62
2000′		71	69	Dry-adiabatic lapse rate	60
4000′	Condensation level	62	58	Dry-adiabatic lapse rate	58
6000′	Level of free convection	53	53	Moist-adiabatic lapse rate	53
8000′		44	48	Moist-adiabatic lapse rate	48

LAYER STABILITY

Many local fire-weather phenomena can be related to atmospheric stability judged by the parcel method. Equally important, however, are weather changes that occur when whole layers of the atmosphere of some measurable depth and of considerable horizontal extent are raised or lowered. Here again, it is necessary to employ some assumptions with respect to conservation of mass and energy, and the assumption that the adiabatic processes still apply. However, it is often possible to employ these concepts with somewhat greater confidence here than in the case of parcel-stability analyses. Let us first examine how the stability of an air layer changes internally as the layer is lifted or lowered.

When an entire layer of stable air is **lifted** it becomes increasingly **less stable.** The layer stretches vertically as it is lifted, with the top rising farther and cooling more than the bottom. If no part of the layer reaches condensation, the stable layer will eventually become dry-adiabatic. Let us consider an example:

We will begin with a layer extending from 6,000 to 8,000 feet with a lapse rate of 3.5°F. per 1,000 feet, and raise it until its base is at 17,000 feet. Because of the vertical stretching upon reaching lower pressures, the layer would be about 3,000 feet deep at its new altitude and the top would be at 20,000 feet. If the air in the layer remained unsaturated, its temperature would have decreased at the dry-adiabatic rate. The temperature of the top of the layer would have decreased 5.5 × 12, or 66°F. The temperature of the bottom of the layer would have decreased 5.5 × 11, or 60.5°F. Originally, the difference between the bottom and top was 7°F., but after lifting it would be 66 − 60.5 = 5.5°F. greater, or 12.5°F. Whereas the original lapse rate was 3.5°F. per 1,000 feet, it is 12.5 ÷ 3, or 4.2°F. per 1,000 feet after lifting. The layer has become less stable.

Occasionally, the bottom of a layer of air being lifted is more moist than the top and reaches its condensation level early in the lift-

A lifted layer of air stretches vertically, with the top rising farther and cooling more than the bottom. If the layer is initially stable, it becomes increasingly less stable as it is lifted. Similarly, a subsiding layer becomes more stable. **Figure 4-19.**

ing. Cooling of the bottom takes place at the slower moist-adiabatic rate, while the top continues to cool at the dry-adiabatic rate. The layer then becomes increasingly less stable at a rate faster than if condensation had not taken place.

A **descending** (subsiding) layer of stable air becomes **more stable** as it lowers. The layer compresses, with the top sinking more and warming more than the bottom. The adiabatic processes involved are just the opposite of those that apply to rising air.

Since the lapse rate of the atmosphere is normally stable, there must be some processes by which air parcels or layers are lifted in spite of the resistance to lifting provided by the atmosphere. We will consider several such processes.

d. moisture

Thunderstorms can occur at any time of the year, but they are more likely during the warmer half of the year. The dewpoints are higher then. Relative humidities in the lower levels of the troposphere should be as high as possible...at least 50%, so that rising air can carry with it an abundance of moisture to be subsequently released as clouds and precipitation. Spring dewpoints in mid-latitudes should be in the 40s and 50s or higher, and late spring through summer dewpoints should be in the 50s and 60s or higher, and preferably about 70 degrees Fahrenheit or higher to provide ample moisture for thunderstorms.

Sometimes it is relatively dry above the surface but moist somewhat above the surface, typically a few thousand feet up, and sometimes it is stable near the surface but unstable several thousand feet aloft. In such cases, thunderstorms can form above the stable layer. This sometimes occurs in the western high plains, resulting in "mid-based thunderstorms", so-called because their cloud bases are in the range of "mid-level clouds", several thousand feet up. Some of the precipitation from mid-based thunderstorms evaporates before reaching the ground. Precipitation falling from the clouds but evaporating before reaching the ground is called virga. The occurrence of virga under a thunderstorm may indicate an environment favorable for microbursts of very high winds (discussed later). When thunderstorm bases are high (say 8,000 to 12,000 feet above the ground) and are above a dry, stable layer, then as impulses of air rise buoyantly through the unstable air aloft, some air bubbles may receive a push downward from that level and would be colder than the environment, so would accelerate downward, reaching the ground as a microburst.

The important roles played by water vapor distribution and transport will also appear in various discussions later in this book.

e. lack of a mid-tropospheric cap

When cumulus clouds start growing into towering cumulus clouds, but subsequently stop growing at say 8,000 or 10,000 or 12,000 feet up, then the rising air parcels have moved into a stable area. As discussed earlier in this book, when the environment through which the parcels are passing is warmer than the temperature of the parcels, then the rising parcels are CAPPED out at that level. A cap or lid on any further growth exists. The "bottom line" is this: if it is too warm aloft, then thunderstorm development will be inhibited. A good forecast rule-of-thumb based on decades of empirical use is this (for mid-latitudes during the warmer half of the year): if the 700 millibar (hectoPascal) temperature is 12°Celsius (54° Fahrenheit) or higher, then this is a cap which will prevent thunderstorm growth most of the time, and a 14°C (57°F) temperature at 700 mb virtually assures a strong enough cap to stop thunderstorm development. A "cap inversion" occurs in a high pressure system. Highs have slowly sinking, diverging air from their centers to generally eastward (in the Northern Hemisphere), but sometimes a section of the sinking air aloft sinks more persistently than the rest, and since these air parcels warm up as they subside, the environment is gradually warmed. This phenomenon typically occurs anywhere between about the 600 and 800 millibar levels, typically around 700 mb. (The 700 millibar level is the height [about 10,000 feet elevation] to which we must rise to have the atmospheric pressure be 700 millibars, or about 21" on the barometer. The average sea-level pressure is 1013.25 millibars.)

HOWEVER, if the instability is so intense below the cap that bubbles of rising air are struggling to force themselves up, then eventually after a couple or a few hours of convective streams surging into the cap, the capped air will mix with the rising air to cool down the cap. If it cools sufficiently, then the rising air parcels will shoot through the cap --the cap is broken-- and thunderstorm development will be explosive.

Often a shallow cap aloft that is about 50 mb thick and only about 2 Celsius degrees warmer than instability, will set the stage for severe thunderstorms that form as the cap is suddenly broken.

* * * * * * * * * * * * *

Reviewing, then, the necessary conditions that must exist simultaneously over a local area for thunderstorms to occur:

we need LIFT, INSTABILITY, MOISTURE and LACK OF A CAP AROUND 10,000 FEET OFF THE GROUND.

Which of these four conditions is the most important? That is a difficult question to answer, because if any one of these conditions or "ingredients" is missing, there will be no thunderstorm formation! It is analogous to having a certain type of table that requires four legs to remain standing; when one of the legs is missing, the table topples over; thus, which of the legs is the most important?

Chapter 5. THUNDERSTORM LIFE-CYCLE

The life-cycle of a single-cell thunderstorm is rather straight-forward. The cycle can be described in three stages:
THE CUMULUS STAGE
THE MATURE STAGE
THE DISSIPATING STAGE

For most ordinary thunderstorms, their life cycles last from about 20 to 60 minutes. When storms persist for much longer, they are either supercells, which usually last for from one to three hours and produce most of the tornadoes, or they are clusters of thunderstorms. In the clusters, we find individual cells forming and dying alongside other cells in various stages of their individual life-cycles.

Before discussing each stage of a thunderstorm, let us first look at n overview of all three stages:
The cumulus stage (the initial building-up stage) is characterized by updrafts. Cumulus clouds grow into towering cumulus. This is not typically a dangerous stage. You may see a few bolts of lightning and hear some rumbles of thunder and rain may begin to fall as the towering cumulus continue to grow. During this stage, the initial cumulus clouds grow into clouds of significant vertical extent, known as towering cumulus clouds.

The mature stage is characterized by continued updrafts and growing into cumulonimbus, and also by downdrafts caused by the drag of air by precipitation as it falls. In severe thunderstorms, flash flooding rains, hail and downbursts of air (as well as concentrated downbursts called microbursts) may occur, and on occasion, tornadoes form. You can tell whether the damage caused by thunderstorm winds is from a microburst or tornado, since microbursts (and downbursts) blow objects over in one direction, whereas the tornado damage is more circular. In mid-latitudes during the warmer half of the year, showers become thunderstorms (i.e., the first lightning and thunder are observed) when the cloud tops reach a height of about 25,000 feet above the ground. When the convection is more at an angle rather than vertical or nearly vertical, which happens typically during the colder half of the year, there still would be the threshold value of updraft and cloud growth of some 25,000 feet for the showers to become thunderstorms, although in this scenario we are experiencing what is called SLANTWISE CONVECTION, since the growth is on a significant slant rather than straight up.

The dissipating stage is characterized by downdrafts, as the function of the thunderstorm (which is to overturn the atmosphere in the local environment) becomes completed and the air stabilizes. The rain slackens, the winds diminish and clouds gradually evaporate in the sinking air. "Debris" or left-over cloudiness may persist for hours afterwards.

THE THREE STAGES OF A THUNDERSTORM:

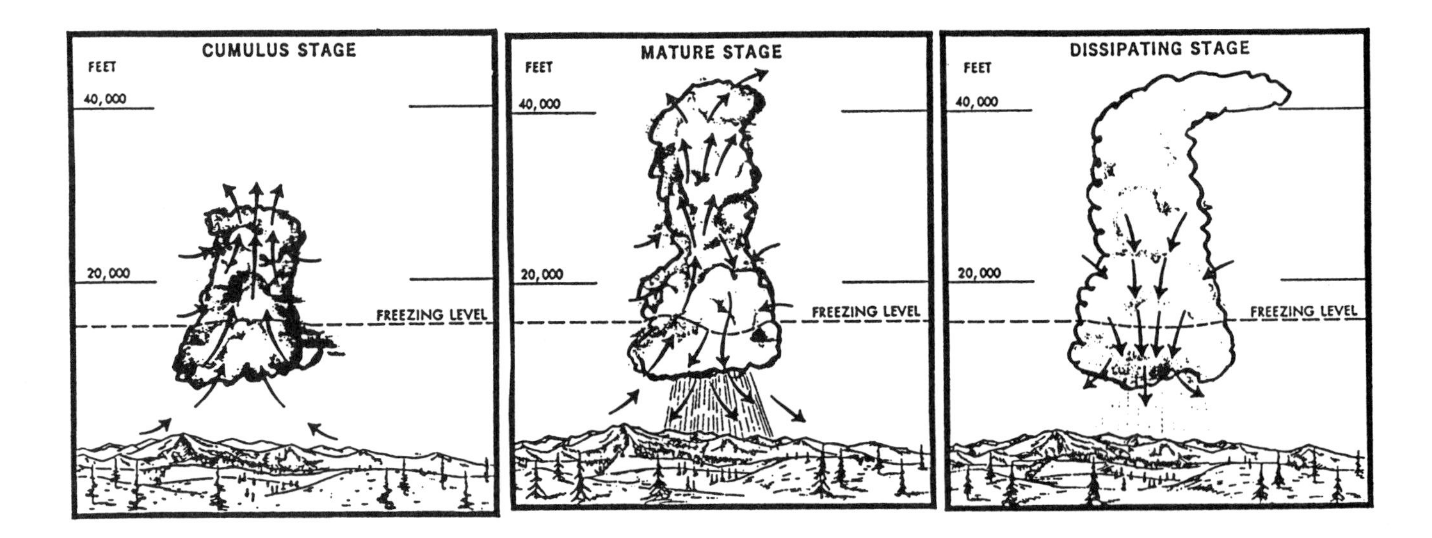

Figure 5-1. The cumulus (left), mature (middle) and dissipating (right) stages of a thunderstorm. In the incipient or cumulus stage, air is converging into the developing cloud from all around, resulting in updrafts. The cloud grows into the mature stage, which is the most active part of the thunderstorm life-cycle. The first rain typically occurs when the first lightning and thunder are observed. The rain and/or other precipitation that would now be falling, drag down colder air as they fall, initiating downdrafts. In the mature stage of a thunderstorm, we find warm updrafts and cold downdrafts. In the dissipating stage of a thunderstorm. Downdrafts characterize this stage as the thunderstorm dies. The precipitation becomes progressively lighter, finally ending, and the cloud matter that remains is debris cloudiness, which eventually dissipates. (source: DOA)

Sample problem:
A thunderstorm is moving from west to east through these cities: Leavenworth (in northeast Kansas), Kearney and Excelsior Springs (both in northwest Missouri). Leavenworth has growing puffy clouds and starts to receive raindrops and some lightning and rumbles of thunder. Thirty minutes later, Kearney is getting hail of 2-inches diameter, very heavy rain, wind gusts over 60 miles per hour and nearly continuous lightning and thunder. Twenty minutes later, Excelsior Springs is receiving light rain, some lightning and thunder, and the tops of the puffy clouds are diminishing in height. What thunderstorm stage is occurring at each city?
Solution:
at Leavenworth: the cumulus stage;
at Kearney: the mature stage;
at Excelsior Springs: the dissipating stage.

Now let us explore the fascinating development of a thunderstorm.

a. the cumulus stage

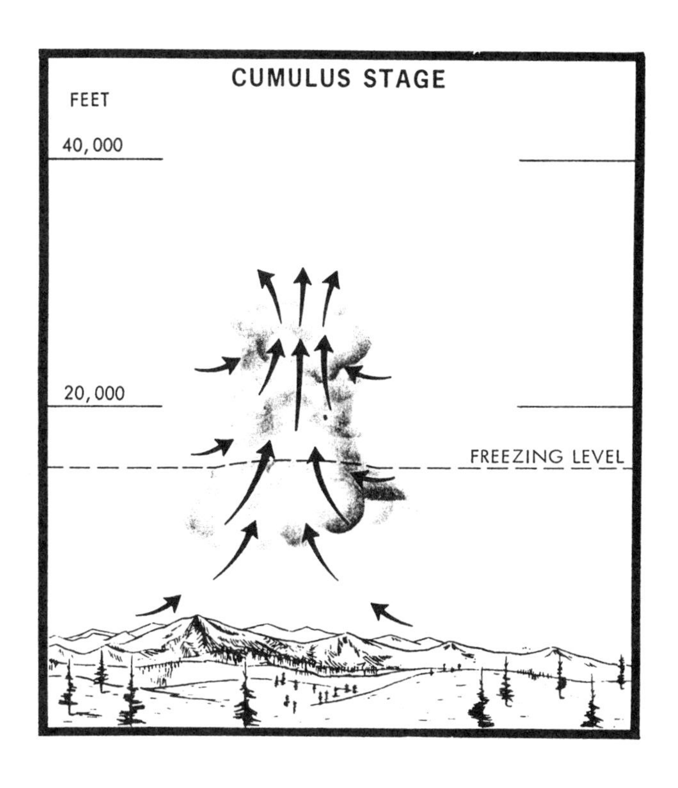

Figure 5-2. The cumulus stage of thunderstorm development.
(source: DOA)

As the puffy cumulus cloud grows vertically, it eventually becomes a towering cumulus, which begins producing showers when its top grows to about 15,000 to 20,000 feet high. In the middle latitudes during the warmer half of the year, the showers become thunderstorms when the tops reach about 25,000 feet. (The exception is slantwise convection, which is discussed in chapter 12.) Thus, the first lightning and thunder occur when the top of the towering cumulus reaches about 25,000 feet high, at which point the cloud is entering the mature stage.

The cumulus stage is characterized by air converging in from different directions, notably in the low through middle levels of the cloud. Since these moist air parcels are piling up, they rise. Thus, the cumulus stage is also characterized by a strong UPDRAFT. The precipitation starts as the cloud keeps growing into the mature stage.

The cellular convection characterized by the strong updraft need not necessarily start at the surface; it can originate somewhat higher, even several thousand feet above the surface. What you as the observer see is a cumulus cloud that keeps growing vertically and also typically is enlarging horizontally.

The speed of the updraft varies spatially and temporally. It often exceeds 25 meters per second, which is over 50 miles per hour, and typically becomes more intense as the storm goes through the mature stage of its life-cycle. The updraft speed is strongest in the center of the cell, weakest along the cell's edges. The rising air is accelerating; thus, the updraft is stronger the higher we go in the cumulus cloud, being strongest around the top of the cloud in this stage of development.

During the cumulus stage of a single-cell thunderstorm development, the air immediately surrounding the growing cloud is slowly sinking over a larger area than that of the updraft.

Initially the cloud droplets are quite small, but grow to raindrop size as this stage progresses. They are carried high into the cloud by the updraft, beyond the

freezing level, where they remain in liquid phase at temperatures that are subfreezing. At the higher levels, the raindrops mix with ice crystals, and at the highest levels only are there only ice crystals or ice particles. During this stage, the raindrops and ice crystals do not fall towards the ground, but are suspended or carried upwards by the updraft.

The temperature of the air within the cloud during the cumulus stage is higher than that of the surrounding air.

Figure 5-3. CUMULUS CLOUDS. When these puffy cauliflower-looking clouds are not growing vertically, then the air is not sufficiently unstable for convective growth, or the moisture or minimal, or there is a cap of warm air aloft which stabilizes the region immediately above the clouds. When the conditions exist for thunderstorm development, the cumulus clouds keep growing upwards, becoming what is known as towering cumulus clouds (next figure). (source: DOA)

As this cell grows into an eventual thunderstorm, keep in mind that most thunderstorms are comprised of several to many cells, each of which is in its own stage of the thunderstorm life-cycle. However, the thunderstorm starts with the first cumulus cell growing vertically.

Figure 5-4. TOWERING CUMULUS CLOUDS. If cumulus clouds keep growing vertically, then they become towers of cumulus called towering cumulus. In mid-latitudes, during the warmer half of the year, they typically produce the first lightning and thunder and become cumulonimbus cloud when their tops reach or exceed about 25,000 feet above the ground. The rising moist air parcels are accelerating upwards during this stage. (source: DOT)

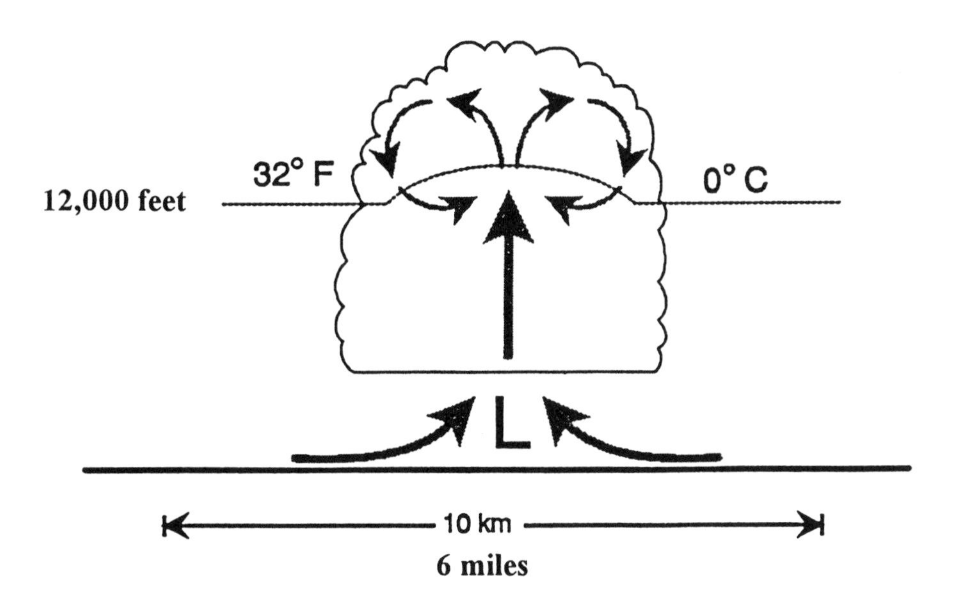

Figure 5-5. Air-flow, as depicted by the arrows, in the cumulus stage of thunderstorm development. There is relatively lower pressure, a weak localized "low", underneath the cloud base due to converging and rising air, lowering the surface pressure. (source: adapted from NOAA)

The towering cumulus stage is the stage between cumulus and cumulonimbus, which is the thunderstorm cloud.

The stage is a time of air convergence that typically is concentrated in a shallow layer of air flowing into the cloud, and is referred to as the INFLOW LAYER. Thus, for the area underneath and immediately surrounding the developing towering cumulus cloud, the wind field shows a gradual inturning of winds forming the area of convergence.

Also, if the cloud is forming during the daytime, the temperature in the shade immediately under the cloud may be up to a few degrees cooler than in the sunny or mostly sunny regions just outside the cloud.

b. the mature stage

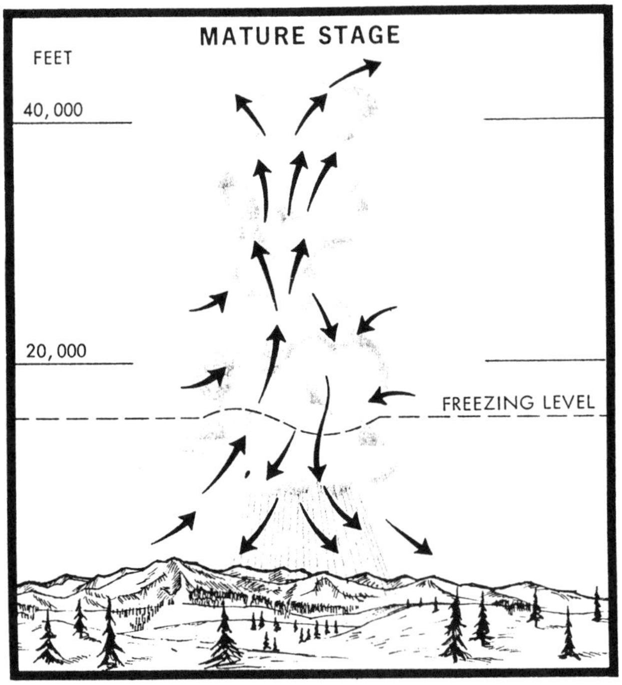

Figure 5-6. The mature stage of a thunderstorm, which is when the most intense and sometimes severe weather occurs. (source: DOA)

When we consider that the average life-span of a single-cell thunderstorm is from about 30 to 60 minutes, then the mature stage exists for no more than one-third to about one-half of that time. Again, we are discussing a solitary thunderstorm cell, not a system comprised of many cells, each of which is in its own stage of development. In a thunderstorm complex of many storms, the entire SYSTEM persists for hours.

The mature stage, which is the most active portion of the thunderstorm cycle, begins when the rain starts falling out of the base of the cloud and the first lightning and thunder begin. Actual showers may start at the end of the cumulus stage, but the clouds will not have grown to the extent which yields lightning; thus, the shower becomes a thunderstorm and the mature stage commences only when the first lightning occurs.

As the rain or other precipitation falls, each precipitation element drags with it a little air. The combined effect of all the precipitation particles dragging down air generates a DOWNDRAFT of air. The mature stage is characterized by the existence of both an UPDRAFT and a DOWNDRAFT. The updrafting air is warmer than the air surrounding the cell, and the downdrafting air is cooler than the air surrounding the cell.

In arid conditions or with high-base thunderstorms and dry air beneath the cloudbase, it is possible to have a thunderstorm with its precipitation evaporating before reaching the ground, which is called VIRGA. The rain or other precipitation is falling into dry air which evaporates it. However, in time, the sub-cloud layer may moisten enough from the evaporated precipitation to then allow the precip. to reach the ground.

Figure 5-7. The cumulo-nimbus cloud of a thunderstorm. This cloud is the mature stage in the thunderstorm life-cycle. (source: DOA)

The precipitation commences when the raindrops and ice particles have grown large enough so that the updraft can no longer suspend them. Thus, large hail implies powerful updrafts, as we shall discuss later. In general, the precipitation starts falling about ten to fifteen minutes after the cell has grown upwards above the freezing level.

It is during the mature stage that the convective cloud reaches its greatest height. The air stops rising at that level. The updraft itself is comprised of parcels of air (from about a square foot to a square meter in volume) that accelerate upwards, reaching a peak velocity usually somewhere just above the middle of the cloud, and then decelerate as they rise above that level until they stop rising. In the strongest of updrafts, the tops of the cumulonimbus clouds reach into the lower part of the stable stratosphere. Cirrus and cirrostratus anvils, often swept downstream by strong winds aloft, often top the cumulonimbus cloud.

Since the top of the troposphere, called the tropopause, is lower the farther poleward we go, we find that the tops of thunderstorms are highest in the tropics and lowest in high latitudes. During the warmer half of the year, thunderstorm cloud tops in mid-latitudes range from 25,000 feet to sometimes over 60,000 feet; in the polar regions, where such storms are relatively infrequent, these tops are usually below 40,000 feet; and in the tropics, thunderstorm cloud tops can exceed 65,000 feet above the surface.

There are several types of thunderstorms and thunderstorm systems, from the ordinary thunderstorm to the supercell to the multi-cell systems, for example. Below is a sketch of the mature stage of an ordinary thunderstorm.

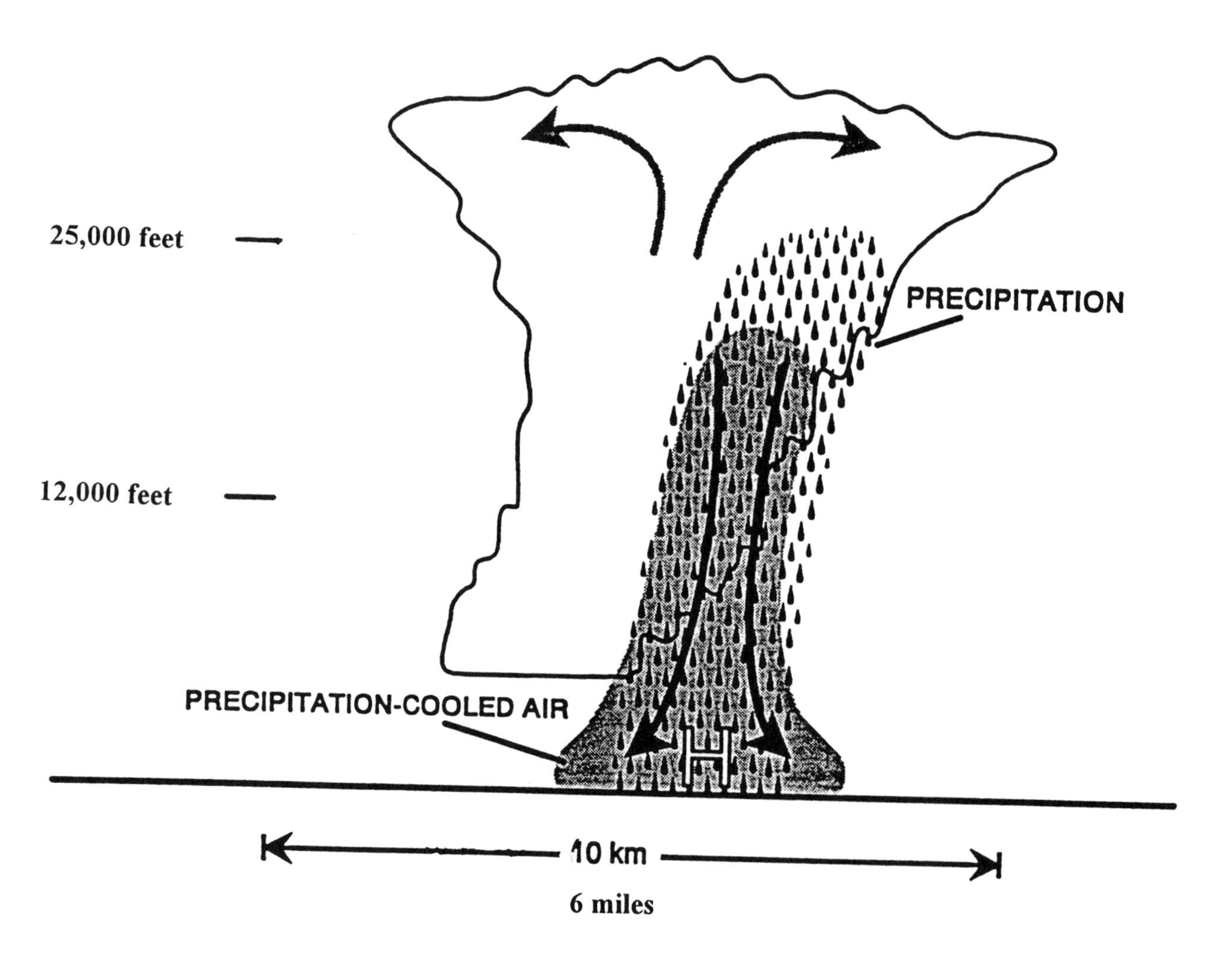

Figure 5-8. The mature stage of an ordinary thunderstorm, showing the updraft and downdraft, the precipitation (shaded region) and a small area of higher pressure (H) in the rain-cooled air at the surface. (source: NOAA)

Recent research and computer models show that the speed of the updraft in an ordinary thunderstorm is from about 20 to 50 miles per hour, but when they exceed that value, the storms are more likely to become severe. As we shall see later, updrafts in excess of 100 miles per hour (over 160 kilometers per hour) are associated with the largest of hailstones. The downdraft speed in an ordinary thunderstorm if from about 10 to 30 miles per hour, but in severe thunderstorms the downdraft speed can exceed 50 miles per hour, even exceeding 100 miles per hour.

The topic of updrafts and downdrafts is quite important, and is therefore covered in-depth in chapter 9.

Above the thunderstorm, the updrafting air, which decelerates eventually to zero vertical velocity, diverges away from the thunderstorm.

Thus, the thunderstorm is a dynamic process, whose mature stage consists of a cumulonimbus cloud with a typical cirriform anvil on top, an updraft, a downdraft in the precipitation, and outflow of cool air from the rain area, and lightning and thunder; sometimes the storm produces hail, and severe thunderstorms produce large hail (three-quarter-inch [2-centimeter] or larger, violent downbursts or microbursts of air (which can exceed 100 miles per hour [160 km/hr]) and sometimes one or more tornadoes.

We have been discussing a single-cell ordinary thunderstorm. Most of the time, we find several cells forming an area of thunderstorms, with each cell having its own updraft and downdraft and each cell being in its own stage of the thunderstorm life-cycle. Thus, we find, for example, an area of say ten cells, five of which are in the mature stage, 2 of which are in the cumulus stage and 3 of which are in the dissipating stage. **To have just one cumulonimbus cloud cell producing a thunderstorm all by itself does happen, but most often we find a number of cells forming a cluster of storms.**

Thus, when Doppler radars detect a thunderstorm area with several updrafts and downdrafts, the radars are showing an area of several thunderstorms, each of which is in its own stage of the thunderstorm life-cycle.

Figure 5-9. The mature stage of a thunderstorm with an exceptionally well-developed anvil top.
(source: USAF)

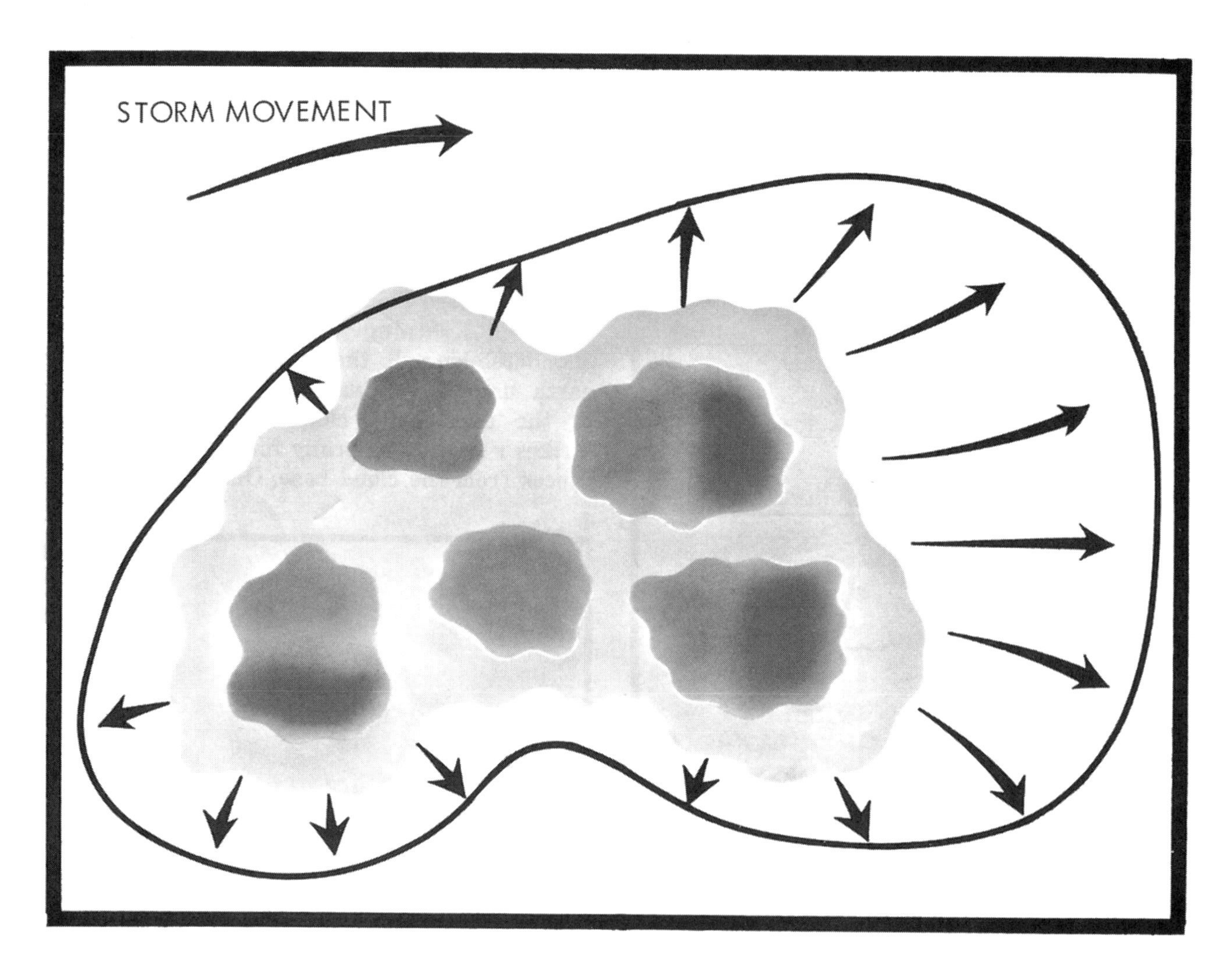

Figure 5-10. The diagram illustrates how a cluster of thunderstorms is comprised of the individual thunderstorm cells which are in various stages of development, and all form one cloud mass. Each pink area illustrates an updraft (cumulus stage of development) and each gray area both an updraft and a downdraft (mature stage of development). Any areas of just downdrafts would be dissipating cells. All of these cells comprise the blob or cluster of thunderstorms. The arrows represent outflowing air...the thunderstorm outflow-boundary...caused by the downdrafts from separate cells, as these downdrafts tend to merge into an outflow of air along the surface, from the thunderstorm cluster. (source: DOA)

Sometimes, in a very moist, warm and unstable environment with no cap aloft, clusters of thunderstorms will merge, forming a large complex of storms known as a mesoscale convective system (MCS), or simply, an large area of organized thunderstorms. These systems (discussed later) can be as big as the state of Iowa, for example, and produce many of our flash floods.

c. the dissipating stage

Figure 5-11. The dissipating stage of a thunderstorm, during which the downdraft eventually spreads over the entire cell, and the updraft disappears. (source: DOA)

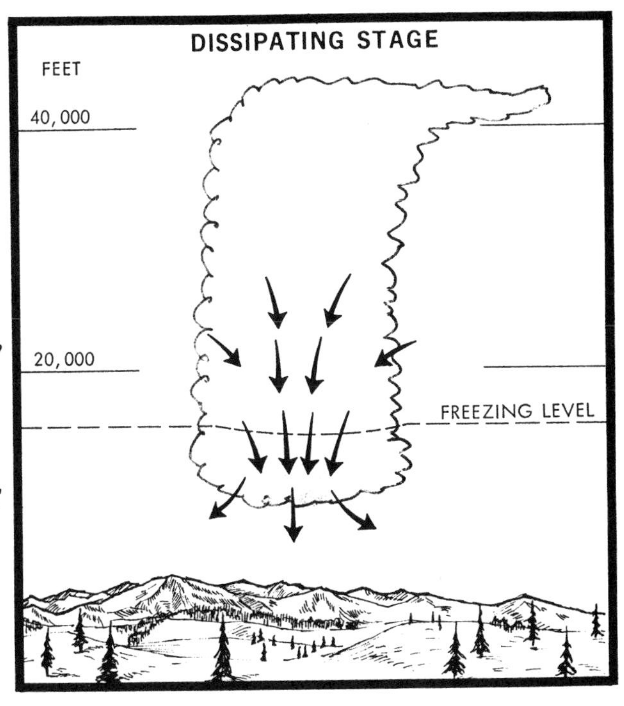

During the dissipating stage of a thunderstorm, the rain becomes lighter and gradually ends. *The thunderstorm has succeeded in its mission, namely, to take a local unstable lower atmosphere and overturn it, stabilizing the local environment.* If redestabilization soon develops, then another thunderstorm will occur if the necessary conditions, as described in the previous chapter, coexist. Thus, although the thunderstorm is certainly a physical "thing", it is also a physical process, which is quite dynamic.

During this final stage of the thunderstorm life-cycle, the downdrafts are continuing to develop and to spread laterally and vertically, as the updrafts weaken. Thus, the cell evolves into a blob of downdrafting air. Since rising air is needed for the development of clouds and subsequent precipitation, and now all we have is sinking air, the precipitation becomes lighter and lighter, then ending, and the clouds dissipate. Actually, some left-over or "debris" clouds may persist for hours, but once the precipitation ends, the storm is considered to be over. Technically, when there is no lightning and thunder, the storm is not a thunderstorm. Thus, the thunderstorm typically ends before the rain shower expends.

The updraft is needed to produce condensation and the release of the latent heat of condensation. In the dissipating stage, since the updraft is no longer occurring, then the source of energy and moisture is cut off. Recall that it is the falling precipitation that drags air down with them to cause the downdraft; thus, when the precipitation ends, the downdraft disappears. During the precipitation, the temperatures within the cell are lower than those of the surrounding local environment, but after the rain ends, the air of the cell mixes with the local surrounding air, and the larger-scale properties of the air take over.

The debris clouds can be at all tropospheric levels, i.e., low, middle and high clouds, or, for example, some low-level stratus may remain for a while, and a detached left-over anvil of cirrostratus and cirrus clouds may also persist for up to an hour or more.

Keep in mind that this discussion is for one thunderstorm cell occurring alone. Most often, cells occur in clusters, and in the more widespread rain events, the clusters merge to form large convective complexes when these clusters form in concentrated zones of warm, moist air. (These zones are called "theta-e ridges", which are important, and are discussed in chapter 22.)

In a cluster of thunderstorms, each cell is in its own stage of development. The downdrafting outflow boundary from one cell can add additional lift to enhance other cell development. Thus, there is a lot going on in interacting thundersnow cell evolution.

As we shall see later, an ordinary thunderstorm, which typically lasts from 20 to 60 minutes, can grow into a severe thunderstorm, such as a supercell, and persist for up to three hours and sometimes longer, when there is significant vertical change in wind direction and/or wind speed in the environment in which the storm is developing or moving into. Too much shear can tear the storm apart, and too little will not enhance severe development. Moreover, the right balance of buoyancy and shear are required for the storm to produce large hail and tornadoes. Also, even without significant vertical wind shear, a thunderstorm in a weakly-sheared environment can interact with some other low-level boundary such as the outflow from another thunderstorm or a sea-breeze front, for example, which can lead to a stronger storm which can produce severe weather.

An ordinary or routine thunderstorm is typically from a few miles to about 10 miles (6 kilometers) or more across. A cluster of storms can range to some 100 miles (about 60 kilometers), and merging clusters that form what is called a mesoscale convective system or MCS for short, can be as big as the state of Iowa or somewhat bigger, or about 300 miles (some 480 kilometers) wide in their widest measurements, and sometimes bigger. Some MCSes are larger than small hurricanes (and occasionally an MCS will move into warm waters such as the Gulf of Mexico and evolve into a tropical cyclone).

The bases of the clouds of most thunderstorms range from just a few hundred feet off the ground to some 8,000 to 15,000 feet above the surface in semi-arid regions such as parts of the southwestern United States. The tops of the clouds are highest in the tropics and are progressively lower as we approach the poles. How high the clouds tops grow to depend chiefly upon the strength of the updrafts, the amount of moisture being lifted and the height of the tropopause, which is where the troposphere ends and the stratosphere begins. Since the tropopause is highest in the tropics and lowest in the polar regions, the highest thunderstorm cloud tops are usually found in the tropics. In mid-latitudes, where most people live, the tops of thunderstorm clouds are highest in the warmer half of the year and lowest in the colder half. The tropopause is highest in the summer and lowest in the winter. In the tropics, these tops can exceed 60,000 feet, while in mid-latitudes in the summer the tops are typically in the 35,000 to 55,000 range, but do exceed 60,000 feet at times. In extreme cases, tops exceed 70,000 feet. In polar regions, a healthy summer thunderstorm can occur in the Fairbanks, Alaska region, for example, with tops of about 25,000 to 35,000 feet.

In the dissipating stage, the rain ending at the surface may not mean that no precipitation is still falling through part of the cloud, since the precipitation can be evaporating before reaching the surface.

Throughout this discussion on the life-cycle of an ordinary thunderstorm, we have considered rising parcels of air. When these parcels are percolating, or rising feverishly en masse, then the convection can be considered to be occurring with a rising layer of air, rather than by parcels, even though in the final analysis, this rising unstable layer is comprised of individual parcels.

d. entrainment

Entrainment is the mixing-in of some of the air surrounding a small weather system such as a thunderstorm, into the system itself.

It is easy to picture how some of the surrounding air about a thunderstorm will mix with the air along the edges of the storm.

For example, consider the updrafting air as the thunderstorm develops. Some of the cooler and usually somewhat drier air surrounding the updraft is entrained into the updraft. A visible manifestation of this entrainment may be some evaporation of some of the external cloudiness.

If an intrusion of enough dry air occurs in mid-levels of the thunderstorm, this destabilizes that region even more, which often results in enhanced convection. Indeed, mid-level dry-air intrusion can cause an ordinary thunderstorm to become a severe one. This type of air intrusion is not considered to be entrainment, but the advection or movement of drier and/or cooler air by generally stronger winds, into the heart of the convective system.

Chapter 6. THUNDERSTORM PROPAGATION AND MOVEMENT

Thunderstorm clusters can grow in certain directions, which seems to give the appearance of movement towards those directions; this is called propagation. At the same time, storms usually are also moving towards some direction. Thus, *the resultant movement of a thunderstorm, cluster of storms or mesoscale convective system is the combination of the actual movement plus propagation.*

Over non-mountainous terrain, observations have shown that, as a first approximation, most individual thunderstorms or small clusters of these storms move with the average wind direction and speed of the approximately 700-millibar (700 hectoPascal) surroundings in which they exist. The 700 mb level air pressure level averages about 10,000 feet in elevation. Over rugged mountains, thunderstorms tend to move with the airflow about 10,000 feet above the surface, but are disoriented due to the influence of variable terrain. For the special case of the mesoscale convective system (MCS), which is a large blob of thunderstorms and rain formed by the merger of several storm clusters, we find that the MCS moves with the 1000-to-500-millibar thermal wind (thickness pattern). As discussed in the technical section in the back of book, the "thickness" is an analysis of lines of equal height difference between two pressure levels. Thus, the 1000-to-500-millibar thickness analysis is a chart showing the lines of equal height difference from the 1000-millibar pressure surface to the 500-millibar pressure surface. The MCS moves parallel to the thickness lines. The figurative "thermal wind" blows parallel to the thickness lines, which is why MCSes are said to move "with the thermal wind." What all of this means is that MCSes tend to stay within the same thermal field in which they formed, if they are to survive as MCSes. The average life-span of an MCS is from 8 to 16 hours. Compare that with the average life span of an ordinary thunderstorm (up to about an hour), and for clusters (which last up to several hours) and for a supercell (which can last for 3 or more hours).

To complicate the forecasting of resultant thunderstorm motion, sometimes a cell will split in two, with one cell continuing as it was (most often, in the Northern Hemisphere, to the northeast, and the second cell heading southeast or even south. This second cell tends to be a severe cell, and is one type of cell that can produce tornadoes.

The thunderstorm is both a thing and a process. It is a thing because it is comprised of matter. It is a process because its composition is continually changing. The movement of the thing is its advection, and the changes due to composition changes is the propagation. Thus, we can meteorologically describe the resultant motion of a thunderstorm or complex of storms as the sum of its advection and propagation.

The figure below shows a blob comprised of cells labelled A, B, C, D, E and F. A short while later, cell A has dissipated and the blob is now comprised of cells B, C, D, E, F and new cell G. Yet a short while after that, cell B has dissipated and the blob is made up of cells C, D, E, F, G and new cell H. Since all cells have been moving with the 700-millibar flow to the east-northeast, the resultant motion is as shown.

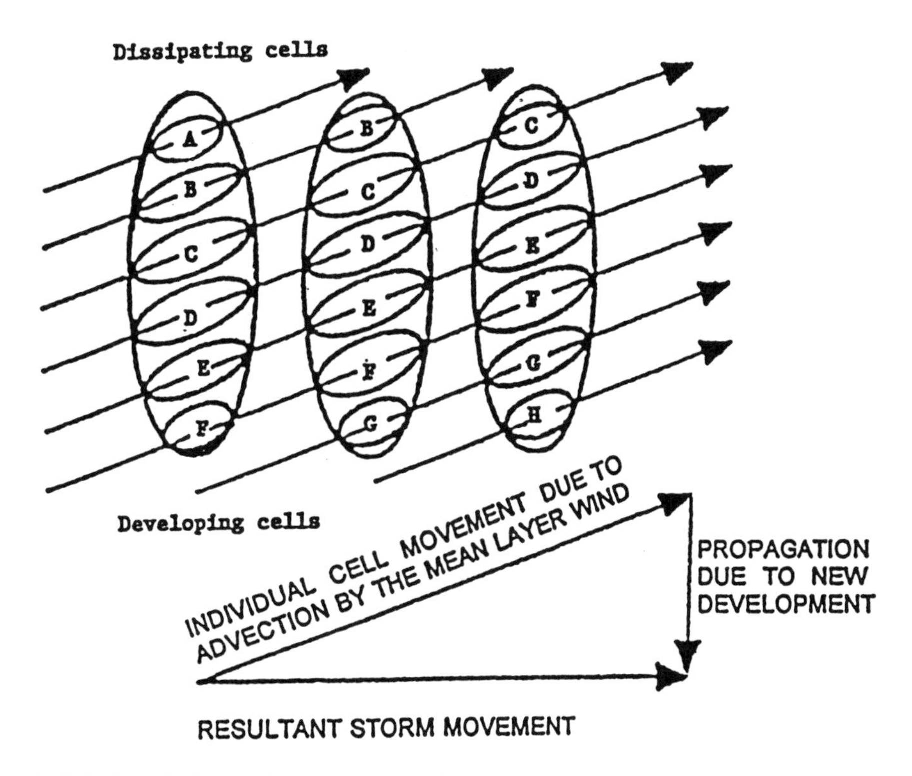

Figure 6-1. A blob of thunderstorm cells moving east-northeastward, but showing the dissipation and generation of individual cells. The top of the page is north and the right is east. In this example, the northernmost cells are dying while new cells form at the southern cell of the blob. (source: NOAA)

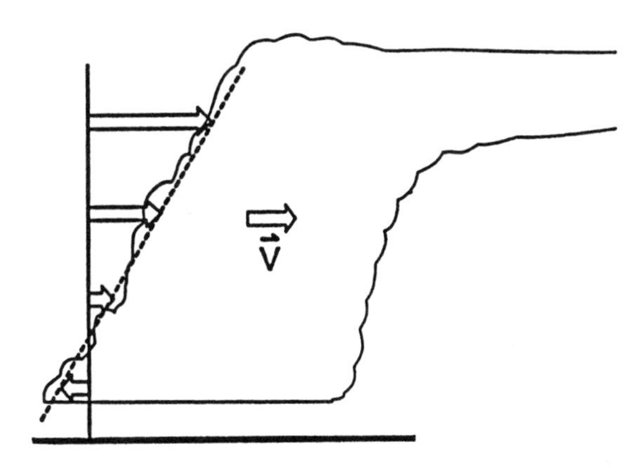

Figure 6-2. The wind shear in the vertical, that is, the change of wind direction and/or wind speed in the vertical, also affects the movement and appearance of the cell or cluster of cells. In this example, the arrows are proportional to the wind speed at each shown elevation. The thunderstorm then takes on a tilted appearance.. The wind and its shear contribute to the movement of the storm as a whole with the vector average shown by the vector V. (source: NOAA)

Figure 6-3. How the storm is tilted often shows the general direction of movement, if you can see the cross-section of the storm. This example is over mountainous terrain. (source: DOA) PAGE 64

Figure 7-1. Electrical discharges from thunderstorms come in the form of streaks of lightning, as shown here, including those streaks that illuminate part of the cloud when they occur inside the cloud), and balls of lightning (rare) known as ball lightning. Electrical discharges also cause optical phenomena that stretch to much higher in the atmosphere, known by such names as red sprites and blue jets.
(source: NOAA)

Figure 7-2. Lightning is a form of electricity that can not be seen by the human eye. This electricity, which is comprised of particles, "bangs" into air atoms, causing them to give off visible light (to glow), therefore allowing our seeing the lightning. (This reason is also why we see the aurora: particles from the sun "bang" into the earth's atmosphere, causing the particles to glow. the earth's magnetic field causes the concentration of the particles to be greater in the vicinity of the magnetic poles.) (source: USAF)

According to contemporary climatological statistics, the earth averages having about 44,000 thunderstorms each day! That is a lot of energy. In your hemisphere, more occur during the warmer half of the year, when convective processes are more active. There are some 9 million bolts of lightning ripping though the air each day, which is a rate of about 6,000 per minute or about 100 every second. A routine thunderstorm produces about 10,000 lightning strikes, and the supercell thunderstorm, which lasts three or more times longer, produces much more.

Some United States Lightning Statistics

•On the average, there are some 26 million lightning strikes annually.
•In the 1940s, when the population was much less than around the year 2000, there typically more than 2000 lightning deaths each year; however, as we approached the year 2000, the annual lightning death was around 100, thanks to weather education about lightning safety. Flash floods kill more people than does lightning, and most flash floods also come from thunderstorms. Thus, flash flood safety education is also a priority. One chief way people die in flash floods is by driving a vehicle into water that is overflowing from a stream, which can overturn the vehicle.
•Thousands of fires are started every year by lightning strikes. Especially vulnerable are forests in the American west where summertime precipitation is mostly light and infrequent.

A single bolt of lightning can have up to 100,000 volts of electrical potential and up to 20,000 amperes of electric current. It takes only about one-tenth of an ampere (amp) to kill a human being! Lightning also momentarily heats the air it is passing through to as high as about 10,000 degrees Fahrenheit, which is the temperature of the surface of the sun (the interior of the sun is much, much hotter).

Two common misconceptions about lightning:

1. Lightning does not strike the same place twice.

This is false, because tall objects (e.g., the Empire State Building in New York City) may be struck several times during a single thunderstorm.

2. It is the rubber tires of a vehicle, or the rubber soles on shoes, that protect people from being struck by lightning.

This is also false, because lightning can travel many miles through the air, which is an insulator, so it can readily travel through several inches of rubber found on a tire and through less than an inch of rubber found on shoes. It is the metal of the car that safely carries the electricity that strikes it, to the ground.

Figure 7-3. A case of lightning traveling out of the side of the cloud, then arcing down to strike the ground. This is a "Bolt out of the Blue." Thus, kids playing soccer or softball need to stay under protection off the playing field even if the sky directly overhead is blue, when the edge of the thundercloud is within a few miles. (source: USAF)

Types of lightning:

Lightning is categorized by the following descriptions: intercloud lightning, cloud-to-cloud lightning, cloud-to-air lightning, cloud-to-ground lightning, ball lightning and lightning into space from the tops of thunderstorms.

Intercloud lightning, which includes intracloud lightning, is when the lightning bolt occurs within a cloud or clouds, so that the cloud or clouds light up.

When the bolt is visible as it streaks from cloud to cloud, it is categorized as cloud-to-cloud lightning.

When the bolt darts from the cloud outward but does not strike a target, it is cloud-to-air lightning.

When the lightning bolt extends from the cumulonimbus cloud to a target on the ground, it is cloud-to-ground lightning.

When the rare case of lightning occurs in which it is a ball of sparks descending from the storm and rolling along the ground, it is ball lightning. This rare type of lightning, averaging about the size of a basketball, is attracted to (it chases!) anything that is animate. It chases after people and animals and has been documented entering aircraft and even squeezing through keyholes and under doors to enter rooms and resumes its ball shape, then proceeding to chase a person, dog or cat, for example. Ball lightning persists for from seconds to usually no more than about one minute.

Figure 7-4. An illustration showing some different types of lightning strokes. (source: DOA)

Lightning can also shoot upward into space (lightning-to-space form of lightning).

The term "sheet lightning" is sometimes used to describe the condition of part of the thunderstorm cloud lighting up because the lightning bolt is within the cloud. When the bolt is visible, its appearance is also sometimes characterized by terms such as forked, zigzag, streak and chain lightning. As shown in figure 7-3 on the previous page, it is possible for a lightning bolt to come out of the thunderstorm sideways and then dart down, striking a target such as a person a few miles away from the thunderstorm, who may be under sunny blue skies! This phenomenon is called "a bolt out of the blue".

St. Elmo's Fire consists of short streamers of light appearing at the ends of pointed objects, especially on mountains, the tops of observation towers, the wings of aircraft and the tips of spars and masts of sailing vessels. It is created by electrical charges building up on the earth and leaking into the air. St. Elmo's Fire often precedes a thunderstorm. It gets its name from St. Elmo who is the patron saint of sailors.

The most likely candidates for a lightning strike from regular lightning are the tallest objects, such as skyscrapers and towers. In more open areas, trees are potential targets. If you are the tallest object around in a thunderstorm, then you are in danger of being struck.

Besides hearing thunder and looking for lightning visually, another way you can tell if lightning is around or near you is that it creates static on AM radio stations.

How lightning forms:
The current thinking about the electrical charge considerations of the ground and the thunderstorm is as follows. Lightning forms when the difference in electrical charge is great between the ground and the thunderstorm cloud. The two types of electrical charge are positive and negative. The ground is normally negatively charged. When a

thunderstorm passes overhead, the negative charge in the bottom of the cloud induces a positive charge in the ground below the storm. In the cloud vicinity the air ionizes, i.e., either gains or loses electrons, causing a conductive channel over which the lightning flows.

After the path for the lightning has been determined by a developing electrical potential differential between the cloud and the ground, then the lightning starts to follow that path to the target. The initial electricity, before we see the visible lightning, moves very swiftly in increments or steps, as depicted in figure 7-5 below.

The first darting of electricity from the cloud is called the step leader. It is invisible to a human observer. The step leader moves rapidly, at about 450,000 mph, but in spurts...that is, in steps, not continuously. Each step is probably on the order of from 150 feet to 400 feet. The very brief hesitations between the surge steps might be caused by the recharging of the tip of the surge with charged particles from the cloud base.

The step leader typically takes the path of least resistance and can branch out up to several times.

When the main stem of the step leader gets to within about 150 to 200 feet of the ground, an upward streamer of positive ions is triggered. This is called the return stroke. This is able to happen because the electrical potential difference between the negatively-charged step leader and the positively-charged ground target is sufficiently strong to trigger the electrical discharge. Then, when this streamer hits the step leader, the circuit is complete.

Figure 7-5. At left, a step leader moving down in steps from the cloud base; and at right, the first electrical streamer connects with the step leader.

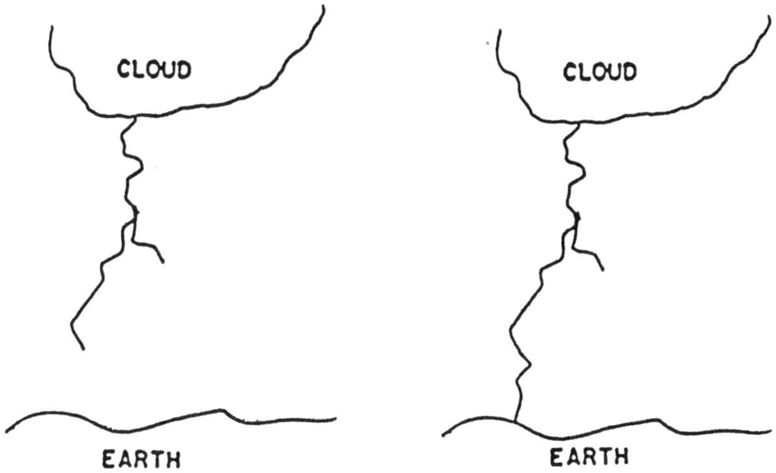

Time-lapse videos of lightning indicate that the brightly luminous return stroke races up the pathway at over 100,000,000 mph. The temperature of a strong electrical discharge can exceed 50,000°F, although commonly, the temperature is about 10,000°F.

Immediately following this upward discharge is a charge shooting downward to the ground as a lightning strike.

All of this is happening so fast that to us as observers it appears as one bolt of electrical discharge.

Sometimes multiple strokes occur from about one-fiftieth to about one-fifth of a second between each return stroke. This is usually long enough for the naked eye to sense flickering.

On the average, about 4 sequences of step leaders and return strokes comprise a single lightning bolt. The typical lightning flash averages 3 to 5 strokes. Some flashes have been observed to have up to 26 separate strokes.

Over 95% of cloud-to-ground lightning strikes are negatively charged. The positively-charged strikes are significantly more powerful and last longer than the negative strokes.

Most deaths from lightning occur in the open, with most fatalities during the afternoon. Most air mass thunderstorms develop during the afternoons and that time is also when the largest number of people are outside. Some people are killed by lightning during the evening and night, with the fewest deaths around daybreak.

Figure 7-6. Zigzag, forked lightning over Swampscott, Massachusetts Harbor.
(photo by Mark Garfinkel)

Some lightning safety information:

•Figure 7-7. If you are outside during thunderstorm activity and your skin starts to tingle with your hair standing straight out, this means that lightning has picked you for a target. You may have typically only about one to two seconds to put your hands on your knees and drop to the ground, making yourself as small a target as possible, so that hopefully the lightning would strike something near you that is a little taller. (source: NOAA)

Figure 7-8. The mother of these two boys took their picture in the western U.S. mountains as their hair stood straight up. From one to two seconds after this picture was taken, both boys were struck by lightning. (source: NOAA)

•Get out of any lake, pond and swimming pool! Non-pure water conducts electricity, and a lightning bolt hitting the water could electrify you.

•Finish your shower and bath immediately and get out of the water. If lightning makes a direct strike on the house and comes through the plumbing, it can electrify you while you are taking a shower or bath!

•Believe it or not, and this is rare: lightning can zap you when you are on the toilet.

•Definitely get to shelter if you hear thunder or see lightning when you are on a field, including being on such fields as baseball/softball and soccer fields. These areas are wide open, so that people would be the tallest objects that would attract lightning. Parents: take your children off the field immediately when there is thunder or lightning.

Even if the lightning is a few miles away, it has been known to dart out of the cloud sideways and then move down to strike a target. Also, get off a golf course (holding a golf iron and playing golf during a thunderstorm is double jeopardy).

Figure 7-9. A vivid example showing lightning's propensity for striking the tallest objects, as in this city. (source: NOAA)

•Get off the telephone. In Kansas City, Missouri two teenage girls in the same neighborhood were killed when lightning hit the phone lines and came through the telephones that they were using.

•Even with one or two surge protectors on appliances such as televisions and computers, a direct lightning strike on or near a house can send the powerful electric surge through the appliance, causing an explosion and possible fire. Therefore, you should unplug appliances such as computers and television sets, especially during storms with excessive lightning.

●Never use an umbrella with a metal handle that is not covered with wood, rubber or some other protective material. Metal conducts electricity.

●You are usually safe in a car. The lightning goes along the car body into the ground.

●During a thunderstorm, never take shelter under a tree. If lightning hits the tree, you could also be struck by the bolt. In some cases, with the most intense bolts, the heat from the lightning may cause tree sap to instantly boil, causing the tree to explode!

Figure 7-10. An observer inspecting the site where a lightning bolt made a direct hit. (source: NOAA)

When lightning strikes a person, much of the lightning goes around the perimeter of the body, and can stop the heart. If cardio-pulmonary resuscitation (CPR) is applied right away, the person can usually be revived.

Lightning can enter your body through your feet after striking the ground hundreds of feet away.

Some people seem to attract lightning. Every now and then you may read about a person who has been struck by lightning several times. There was even such a man who had been blind, and a lightning strike restored his sight (he claimed).

Figure 7-11. Because even the lightning from a routine non-severe thunderstorm is dangerous, and because the intensity and frequency of lightning discharges are related to the rainfall rate and severity of a thunderstorm, the detection of lightning is important. (source: NOAA)

Lightning detection:

1. Besides the obvious visual observing of near and of distant lightning, you can also detect lightning by observing increasing amounts of static on AM radio (NOT on FM).

2. Lightning detection systems are in use, which use a device to detect cloud-to-ground (called CG) lightning strikes. It detects radio waves produced by lightning. These waves are called sferics (derived from the word, "atmospherics"). A network of these devices across the United States and Canada allows scientists to study lightning activity from a storm as it travels. We have learned from this that the most intense part, including the most intense rainfall, from a thunderstorm occurs where the greatest frequency and concentration of CG lightning strikes occurs. The computer monitor displays from lightning detection systems typically show colored dots for each CG strike. The fiercest strikes are the 5% of CGs that are positively charged (about 95% are negatively charged).

Some beneficial aspects of lightning:

1. Lightning helps to balance the electrical charges of the atmosphere with those of the earth. Recall that before the lightning strike, the bottom of the cloud is usually negatively charged and the ground is positively charged. A lightning bolt tends to return electrons from the cloud to the ground, thus making both the bottom of the cloud and the ground neutral in relation to each other.
2. Lightning combines dust, smoke and other solid particles in the air into larger particulate matter that fall or are washed to the ground by the falling rain. Have you

ever noticed how clear the air is after a healthy thunderstorm? Without rain and lightning, the lower atmosphere would become so polluted that life would be threatened.
3. Lightning is the world's largest manufacturer of natural fertilizer. It produces some 100 million tons of nitrogen compounds annually! This is done by the lightning causing nitrogen and oxygen in the air to combine to form <u>nitrates</u>. Nitrates are solids that dissolve rapidly in rainwater, which causes them to fall to the ground. They penetrate the top layer of the soil and are absorbed by plants. Plants use the nitrates to produce plant protein, without which they would die.

Trivia item: the word "electricity": The Greek philosopher, Thales, experimented around 600 B.C. with a piece of the fossil resin, amber, ("elektron" is the Greek word for amber, and the word has similar roots in Latin), rubbing it furiously with a dry cloth, which then allowed the amber to attract feathers or straw. In the 1500's, the court physician to England's Queen Elizabeth, William Gilbert, repeated this experiment and then named the science of studying it "vis electrica".

<u>Ball Lightning</u>:
Imagine sitting one evening in front of the TV, eating snacks and enjoying your leisure time, when suddenly a ball of sparks, about the size of a basketball, bursts into the room and proceeds to head towards you. Startled, you would undoubtedly try to flee the approaching unknown entity. However, this ball would rapidly catch up to you and zap you with a low dose of electricity before it dissipated. This is not science fiction. This phenomenon is a rare form of lightning known as ball lightning.

Meteorologists define lightning in the following terms. When the lightning bolt is within the cloud, so that you do not see the bolt itself but rather observe the cloud section flash, this is called **intercloud lightning** or **intracloud lightning**. When the bolt is visible but lances from cloud to cloud, it is referred to as **cloud-to-cloud lightning**. When the bolt extends from the cloud to the air, it is called **cloud-to-air lightning**, and when it darts from the cloud to a target on the earth's surface, it is known as **cloud-to-ground lightning**. These are the four common lightning designations.

However, when the lightning takes the shape of a ball, about the size of a basketball, and appears to throw out sparks as it moves along, it is known as **ball lightning**.

Ball lightning is typically a shade of orange or off-white. It descends from the thunderstorm and rolls or moves along the surface. It has the interesting property of being attracted to anything animate; i.e., it "chases" any person or animal in motion. Meteorologists occasionally receive reports of "balls of sparks" coming out of thunderstorms and chasing people who were caught outside in the thunderstorm.

Ball lightning can squeeze itself very narrow and enter a house underneath the door. It has also been known to squeeze through a keyhole, enter a room, and subsequently resume its ball shape.

Then it would chase a person, dog or cat around the room until it either catches up with him, her or it, or fizzles out.

Figure 7-12. A representation of ball lightning moving along the ground, chasing an animate creature.

One of the strangest stories about ball lightning of which this author is aware, was reported to me when I was the Meteorologist-in-Charge of the National Weather Service Office at Rochester, New York. While working the day shift during a day with considerable thunderstorm activity, an acquaintance who was a commercial airline pilot came into the office and told me about a ball lightning encounter his aircraft passengers experienced while in descent to the airport through a thunderstorm. A "ball of sparks" about the size of a basketball entered the aircraft apparently through an engine intake, moved into the fuselage, and proceeded to chase a flight attendant up and down the aisle. She was screaming as she tried to outrun the ball lightning. The lightning apparently dissipated quickly before striking her.

A logical inquiry about this phenomenon is, "What happens when ball lightning strikes you?" If it reaches you before dissipating, its effect is similar to what you would experience if you are wearing rubber or rubberlike shoes, slippers or tennis shoes, vigorously rub your feet on a rug or carpet for about 30 seconds, and then tough something metal with your finger.

A spark, commonly referred to as "static electricity", occurs between your finger and the metal object. This spark is not only felt, but is visible, especially if the room is dark.

It has been difficult to find any documentation of any person actually killed by ball lightning, but there are sufficient reports of people experiencing a charge similar to the "static electricity" phenomenon to allow some documentation of this rare type of lightning.

Sometimes this lightning variety will enter a building, move along the floor and up the walls, and even across the ceiling. If a moving person, dog or cat is nearby, then it can dart after the person or animal and "zap" him, her or it unless it dissipates.

This raises an interesting question: "Suppose you stand still when ball lightning is approaching you; would it leave you alone, bypassing you?" We can only speculate.

The ball lightning's life span lasts from a few seconds to perhaps over a minute, and may in some episodes persist longer.

A particularly interesting ball lightning episode occurred during the explosion of Mount St. Helens in Washington state in 1980. Observers reported the following, from about 100 miles southeast of the volcano: "The lightning was in ball form, streaking towards the ground, connected neither with the cloud nor with the ground. It was like a group of balls all going in the same direction." Closer to the volcano, at about 29 miles north of it, an eye-witness reported, "After the cloud passed overhead, lots of lightning started at some 600 to 800 feet in the air, and formed big balls, big as a pickup, and just started rolling across the ground and bouncing." This pickup-truck size ball lightning is the largest ball lightning yet reported. Most episodes of it report that the lightning is smaller, typically about the size of a basketball.

Sometimes the ball lightning is even smaller than basketball size. Consider the following report of a ball lightning incident from Russia. In January 1984, ball lightning (the Russian report referred to as the "visiting fireball") entered a Russian aircraft, "flew above the heads of the stunned passengers", and then left out the tail section. The ball lightning left two holes in the plane.

The report did also refer to the fireball as "ball lightning" and stated that it occurred with thunderstorms in the area.

The Russian news agency reported the following, "Suddenly, at the height of 1200 yards, a fireball about four inches in diameter appeared on the fuselage in front of the crew's cockpit. It disappeared with a deafening noise, but re-emerged several seconds later in the passengers' lounge, after piercing in an uncanny way through the air-tight metal wall.

"The fireball slowly flew about the heads of the stunned passengers. In the tail section of the airliner it divided into two glowing crescents which then joined together again and left the plane almost noiselessly."

While repairing equipment aboard the aircraft, mechanics discovered two holes - one in the front of the fuselage and another in the tail section.

Thus, fascinating documentation of ball lightning exists.

SPRITES and other optical flashes above thunderstorms, extending well up into the local atmosphere:
Another product of thunderstorms is optical phenomena excited by the storms. Vertical, mostly-red-appearing luminosity extending upwards is called a SPRITE. Moreover, blue jet-like appearing bursts of luminosity are sometimes visible over thunderstorms, but they begin miles above the storms and extend upward. These are called JETS. Thus, red (or mostly red) sprites and blue jets are associated with thunderstorms. Green emissions called elves have also been observed.

Under a red sprite is typically a blue threadlike, filament-like luminosity. Although sprites can occur singly, they most often appear in clusters.

The term "sprite" is given to this phenomenon to convey its fleeting behavior.

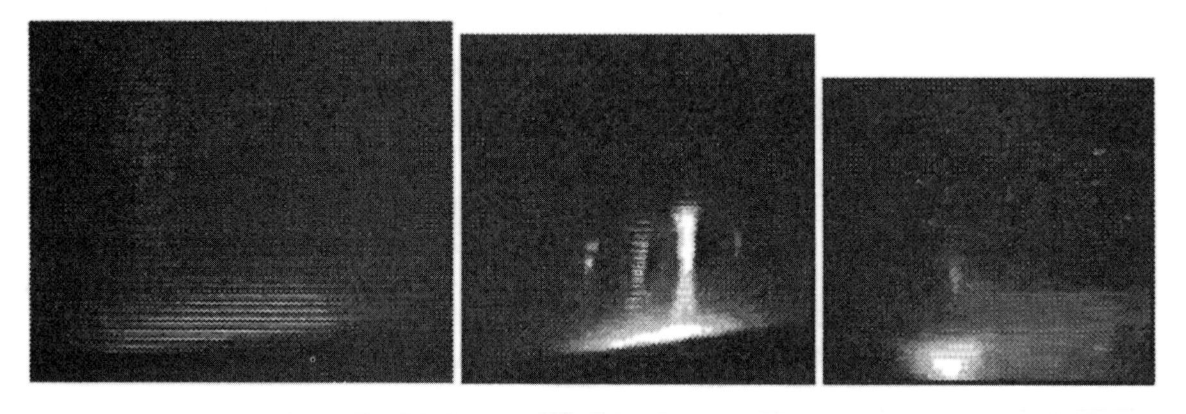

Figure 7-13. Photographs of sprites and jets. (source: University of Alaska)

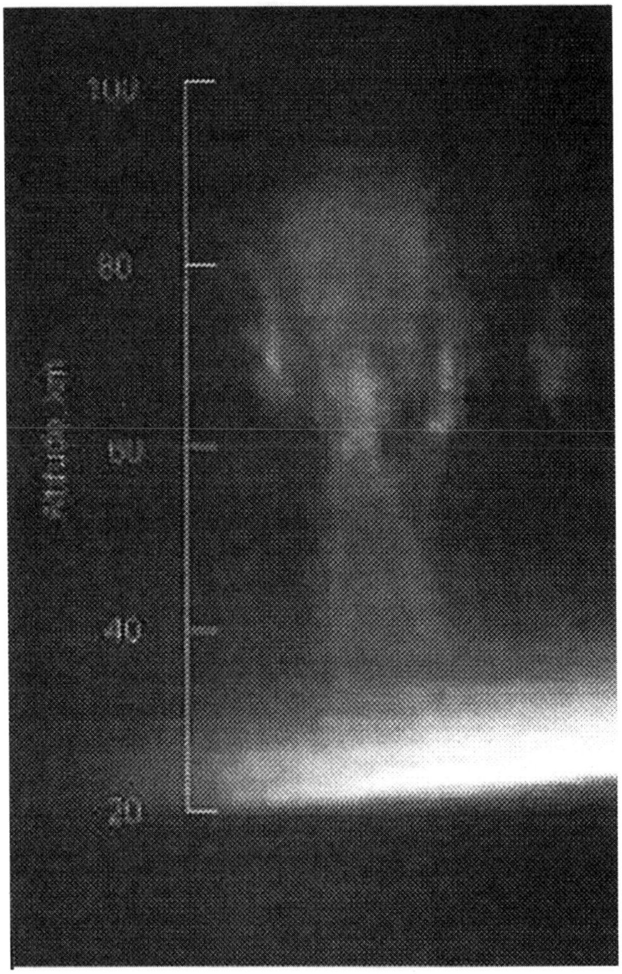

Figure 7-14. Picture of sprites above a thunderstorm. (source: University of Alaska)

Since sprites occur above thunderstorms and clouds obstruct viewing, and since they are seen when viewing against a dark sky background, they are usually difficult to observe. They are typically as bright as moderately-bright aurora. You would need to be away from the light-pollution of the skies by cities to have any opportunity of observing them.

The luminosity of lightning is much brighter than sprites, which also contributes to making sprites harder to see. Sprites appear to occur with only a tiny fraction of all the lightning strokes. Thus, it seems as if lightning must occur in order for sprites to form.

Your best opportunity to observe sprites is at night in the country when thunderstorms are on the horizon. Look above the thunderstorm(s). First, adapt your eyes to the dark night for a few minutes; i.e., be in the dark for a while so that your eye pupils enlarge. This allows you to see fainter lights. You may be able to enhance your chance of seeing sprites if you hold a piece of cardboard or such in front of your view of the storm itself, to block out much of the brightness of the lightning.

These optical phenomena are likely another manifestation of thunderstorm electricity. We can say that sprites, jets and similar optical phenomena are phenomena excited in the middle and upper atmosphere by lightning.

Sprites dance above thunderstorms, which you can compare to aurora shimmering in the sky, Observers in aircraft, including space shuttles, have seen sprites.

Historical note on the origin of the word, "sprite": The term, "sprite' was coined in 1993 over pie and coffee one winter evening in a cabin near Fairbanks, Alaska. This term was chosen rather than a term such as "cloud-to-space lightning". The elusiveness and fleeting appearance of this phenomenon, which lasts about 1/100th of a second or less, led to adopting the word, "sprite".

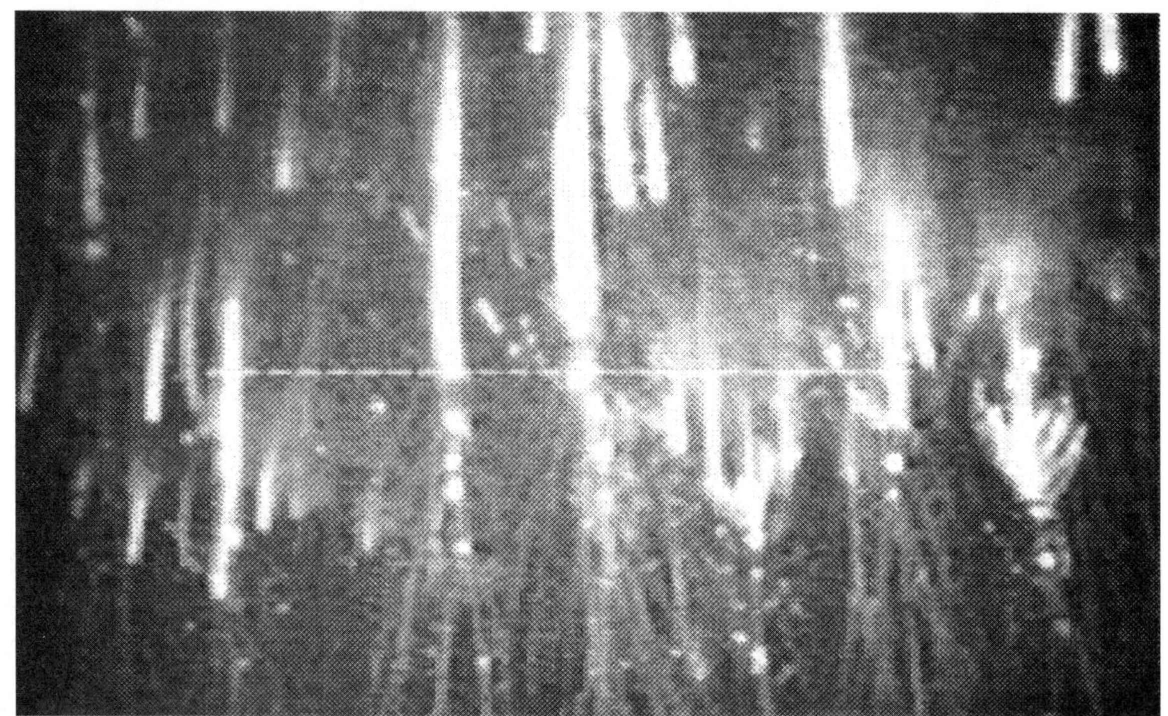

Figure 7-15. Two sequences of images from a video-tape, of dancing sprites above thunderstorms. In the top three sequential photos, we observe a succession of sprites that dance across the screen, moving from right to left. In the lower three photos, taken a few minutes later, we observe a second series of sprites dancing from left to right. The dashed line in these pictures was added by computer to indicate a level from where spectrographic analysis was being done. The tops of these particular sprites were 50 to 55 miles high. (source: University of Alaska)

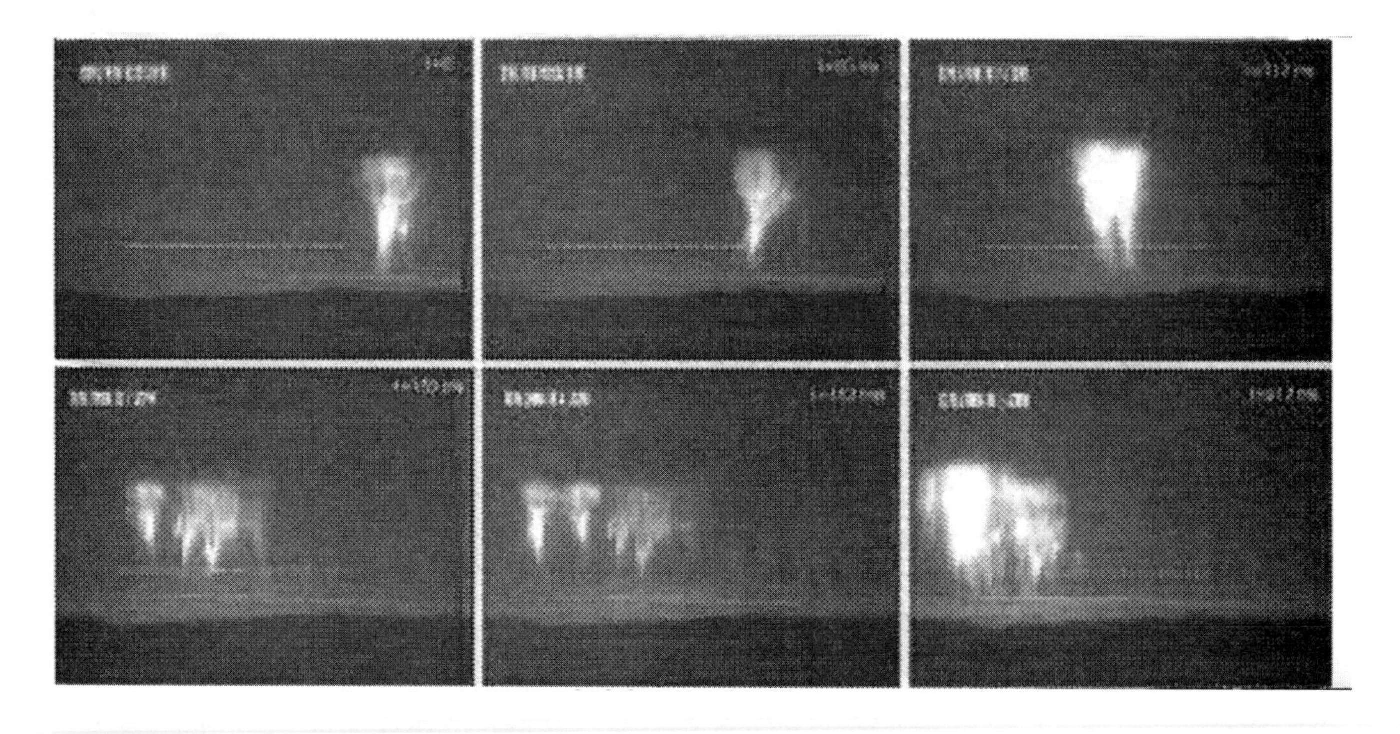

Figure 7-16. A cascade of sprites photographed over thunderstorms near Fort Collins, Colorado. (source: University of Alaska)

Figure 7-17. A picture taken from an American space shuttle, showing lightning in thunderstorms below. (source: NASA)

Figure 7-18. Lightning strikes next to the space shuttle on its launch pad at Cape Canaveral, Florida. (source: NASA)

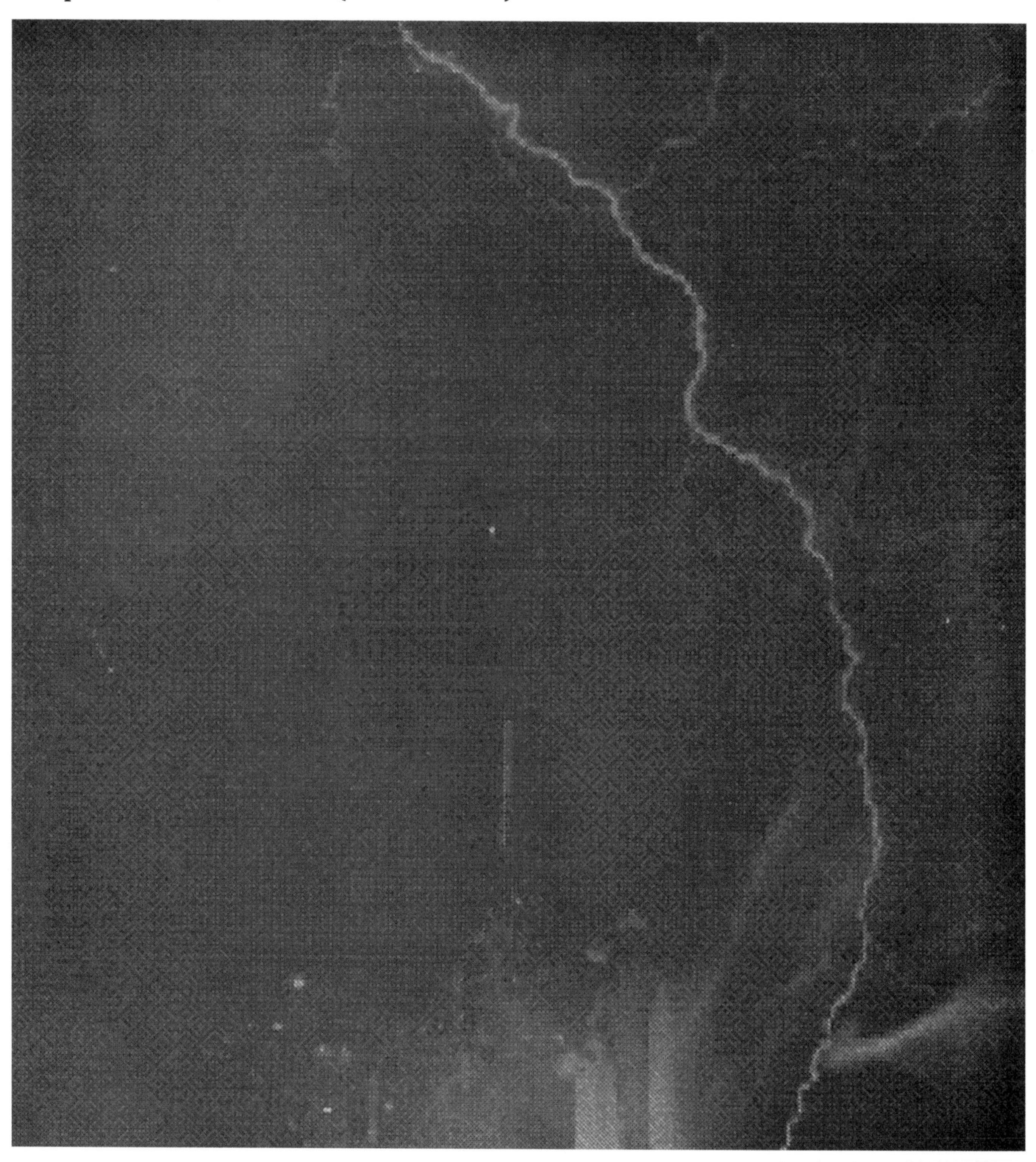

Chapter 8. RAINFALL AND FLASH FLOODS FROM A THUNDERSTORM

The most intense rainfall rates on earth come from convective rains: thunderstorms and tropical cyclones. Even more extreme rainfalls occur when these convective clouds are forced up the sides of a mountain range, which lifting is called orographic lifting. This lifting forces extreme amounts of condensation. Thus, for example, the south slopes of the Himalaya Mountains in India have experienced rainfall totals of over one thousand inches (over 25 thousand millimeters) in one year.

In the United States, a convective cloudburst dropped 12 inches of rain on Holt, Missouri in northwest Missouri on June 22nd, 1947. Other thunderstorm episodes routinely yield amounts in excess of two inches per hour in the United States every convective season. Hurricanes and typhoons have dumped a few feet of rain within 24 hours in the most extreme cases of slowly-moving large tropical cyclones.

Sometimes a stalled out or very slowly-moving front will cause thunderstorms to keep forming and moving over the same area. This is known as training or train-effect since the paths of these storms act as if they are moving along the same train track, thus over the same area.

Some of the most intense snowfall rates occur when thundersnow superimposes itself upon a low pressure system's main snow area. Rates in excess of two inches per hour are common in thundersnow events. Oswego, New York once received eleven inches of Lake Effect snow in one hour, accompanied by thunder and lightning. Although the cloud tops were only around 15,000 feet, the convection, as happens in winter, was on a slant (called slantwise convection), so that if the cloud growth or depth were measured through this slant, the depth exceeded 25,000 feet.

Thus, the most intense precipitation comes from convective weather systems, that is, those weather systems (showers, thunderstorms, tropical cyclones) that are formed by convection, which gives us cumulus, towering cumulus and ultimately cumulonimbus clouds

Most flash floods are caused by convective rains because flash flooding requires intense rainfall over a relatively short time. In mountainous terrain and in highly urbanized areas, rainfall rates of one inch per hour or more can initiate flash flooding

Since the average life-span is a typical or ordinary thunderstorm is usually no more than about an hour, there are typically numerous cells comprising a flash-flooding thunderstorm system.

Figure 8-1. Some of the most intense rainfalls observed in the United States from the late 1880's through the end of the 20th century. All of these rainfalls were convective.

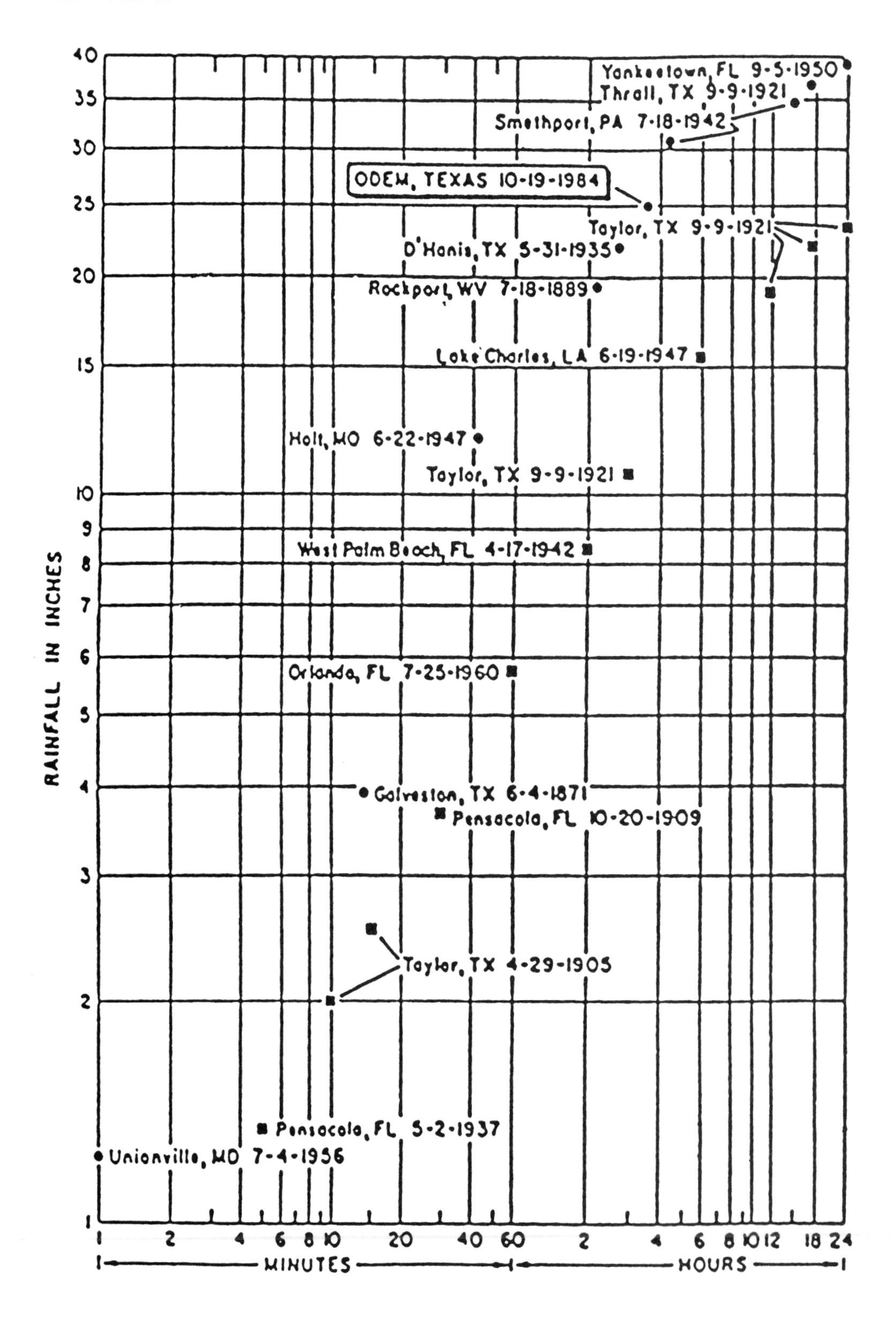

We know that the combination of very high the dewpoints over an area and strong lift can yield significant rainfalls. Another useful tool for anticipating the rainfall potential from thunderstorms expected to develop soon is a weather chart called the **PRECIPITABLE WATER CHART.**

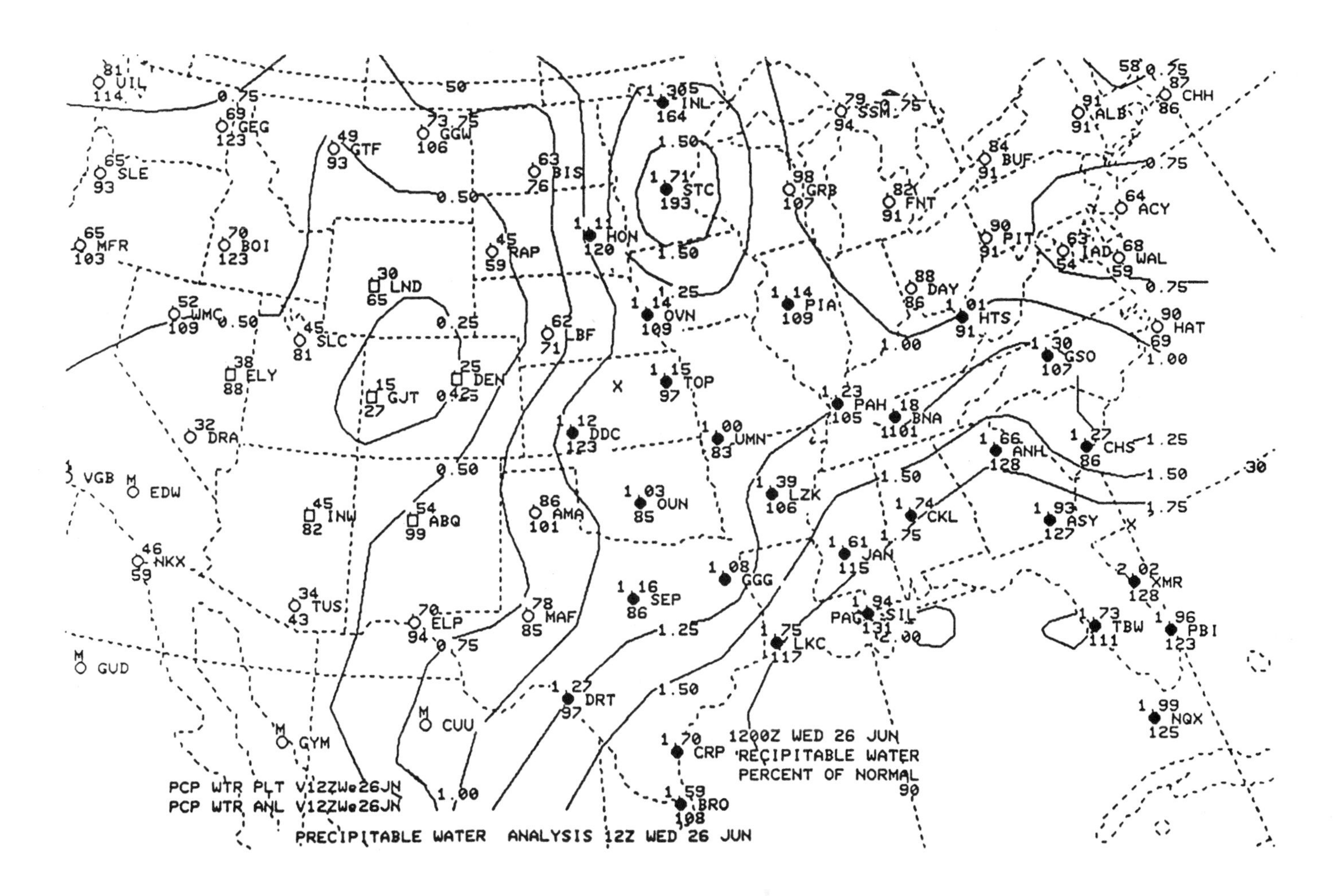

Figure 8-2. A precipitable water (or PW) chart. Precipitable water is the amount of water vapor in a column of air from the surface to 500 millibars. For each location, the PW value is given on top, to the nearest one one-hundredth of an inch, and below the station location circle is the percent of climatological normal for that time. Isolines are drawn for every quarter-inch. As a rule-of-thumb, PW amounts that are about 150% of normal and higher are worth noting, for thunderstorms occurring in these regions can produce excessive rainfall rates.

This chart is developed from dewpoint data gathered by weather balloon radiosondes as they ascent through the troposphere into the lower stratosphere from various sites around the world. Specifically, what is needed for PW determination are data from the surface to the 500 millibar (500 hectoPascal) pressure level (which in mid-latitudes averages about 18,000 feet above mean sea-level). A computer program uses the dewpoint values over each location to compute how much moisture is in that column of air from the surface to 500 millibars. The actual moisture is given in parts per thousand, or grams of water vapor per kilogram of air, and is called the **mixing ratio**. The warmer the air, the more moisture it can hold, so the mixing ratios are highest (several parts per thousand or higher at the surface) in warm air masses. However, the precipitable water is the integration or summing-up of the total moisture in that column from the surface through 500 millibars (500 mb). Since relatively small mixing rations are found above about 500 mb, the integration stops at that level.

The precipitable water chart gives the amount of water in a column of air from the ground to 500 millibars, which means that if all the water vapor above that location could be condensed in precipitation, that is how much would fall out of the clouds. However, that is not how precipitable water values are used. During convection, the low- through mid-tropospheric convergence of air from the surrounding area is bringing in additional water vapor. Decades of observations have shown that during well-organized thunderstorm episodes, the amount of rainfall that can be expected from thunderstorms is typically from three to five times the precipitable water value for that area, and in extreme events can be up to seven times the PW value. Moreover, during train effect thunderstorms, the amounts are even higher.

Look at the PW chart example again. As stated on the previous page, PW amounts that are about 150% of normal and higher are worth noting, because thunderstorms occurring in these regions can produce excessive rainfall rates.

Connection between lightning activity and heaviest rainfall:

In overlaying and analyzing data from lightning detection systems and radar rainfall rate estimates, there is a prominent correlation between where the heaviest thunderstorm rainfall is occurring the greatest lightning frequency and intensity. Recent studies confirm this observation, but also suggest that thunderstorms that form over higher semi-arid terrain that have higher cloud bases, do not always indicate the heaviest rainfall since much of this rain may evaporate before reaching the ground. Nevertheless, for non-high-based thunderstorms, the correlation has been established.

FLASH FLOODING FROM THUNDERSTORMS

As urbanization continues and as the population increases, the number of people perishing in flash floods continues to rise throughout the world. When an area is paved over or blacktopped, the runoff from rainfall is more than if the area were natural which results in some of the runoff going into the soil and vegetation. Thus, flash flood mitigation is a major concern in urban areas, as well as in mountainous areas when the runoff is fast and at times dangerous.

Figure 8-3. Billions of dollars of property damage can occur within one year in the United States alone, due to flash floods from excessive convective rains from thunderstorms and hurricanes, plus other causes of flash floods such as the rapid melting of very heavy snowcover, the failure of a dam or levee, the sudden break-up of an ice or debris jam on a water channel (stream or river) and a major earthquake or volcanic eruption disrupting the flow of water channels. (source: NOAA)

Thunderstorms that produce excessive rainfall are often the result of any or a combination of these factors:

- slowly-moving or stationary heavy thunderstorms
- thunderstorms forming and passing over the same area (train effect)
- merging cells or merging clusters
- a large area of organized convection over the same area for a long time (a mesoscale convective system or a tropical cyclone or decaying tropical cyclone).

Merging cells or merging clusters of thunderstorms are often regions for excessive rainfall since the intersecting of the outflow of winds from these cells or clusters results in significant convergence and enhanced lifting of the air. The result is more intense thunderstorm activity, usually including copious rainfall.

Every year we read and hear news reports of flash floods, including major ones, that cause local devastation and drown people.

Figure 8-4. Raging torrents of water, moving at 30 to 40 miles per hour, rush down the James River at Richmond, Virginia during a flash flood caused by the remains of Hurricane Camille, which dumped up to 30 inches (some 760 millimeters) of rain within a few hours over parts of western Virginia and West Virginia. (Source: Virginia Department of Highways)

Chapter 9. UPDRAFTS, DOWNDRAFTS, DOWNBURSTS, MICROBURSTS AND NOCTURNAL THUNDERSTORM HEAT-BURSTS

Discussions of updrafts and downdrafts in some of the previous chapters introduced this topic. Concerning the life-cycle of a thunderstorm cell, the cumulus stage is characterized by updrafts (or one huge updraft), the mature stage by updrafts and downdrafts and the dissipating stage chiefly by downdrafts.

As discussed earlier, at least one lifting mechanism is required to raise the air to a level at which no additional lift is required, even if it still exists, since from this level the parcels are buoyant (warmer than their environment) and thus keep rising. The level that parcels need to be lifted to in order to become buoyant is called the **LIFTED CONDENSATION LEVEL (or LCL)**. LCLs are typically from a thousand to a few thousand feet above the surface. There are higher LCLs, as high as 8 to 12 thousand feet above the surface, in semi-arid areas such as far western Kansas and other high plains areas such as eastern Colorado and southeast Wyoming, which means that the bases of clouds of such storms are occasionally considerably higher than those of most thunderstorms.

The downdrafts are caused by colder air being dragged earthward by the falling precipitation.

During the cumulus stage of the thunderstorm life-cycle, typical updrafts speeds are on the order of 8 to 15 meters per second, which is about 20 to 35 miles per hour, but as the cell evolves through its mature stage, the updraft strength usually increases to 15 to 25 meters per second, which is about 35 to 60 miles per hour. In the most severe thunderstorms, updrafts can exceed 50 meters per second or about 115 miles per hour. Indeed, grapefruit-size hail requires updrafts on the order of 125 to 150 miles per hour!

The downdrafts are usually not as strong as the updrafts, being typically about 5 to 20 meters per second, or about 10 to 45 miles per hour, during the storm's most intense stage, the mature stage. However, sometimes a downdraft can become more concentrated and can accelerate downwards faster, becoming a downburst of air. A concentrated downburst is known as as MICROBURST, and can exceed 100 miles per hour, causing significant damage.

What separates an ordinary thunderstorm from a severe thunderstorm is the strength of the updrafts. Note that the plural "updrafts" is used because a thunderstorm is, in most cases, comprised of several convective cells, each in its own stage of the thunderstorm life-cycle. Each cell, however, contains one updraft during its cumulus and mature stages.

When a downburst of air is no wider than under about 2½ miles in diameter, it is classified as a MICROBURST.

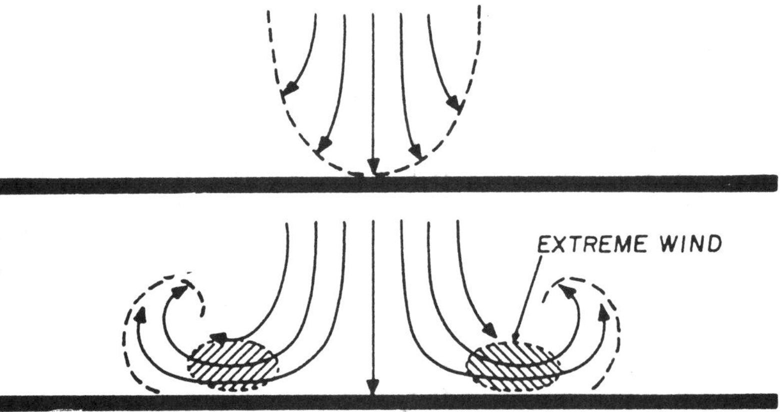

Figure 9-1. A microburst of air accelerating downwards at a high speed, sometimes reaching the ground with a speed of 100 miles per hour (about 160 kilometers per hour) or more (top sketch), and spreading out as it hits the ground (bottom sketch). (source: NWS)

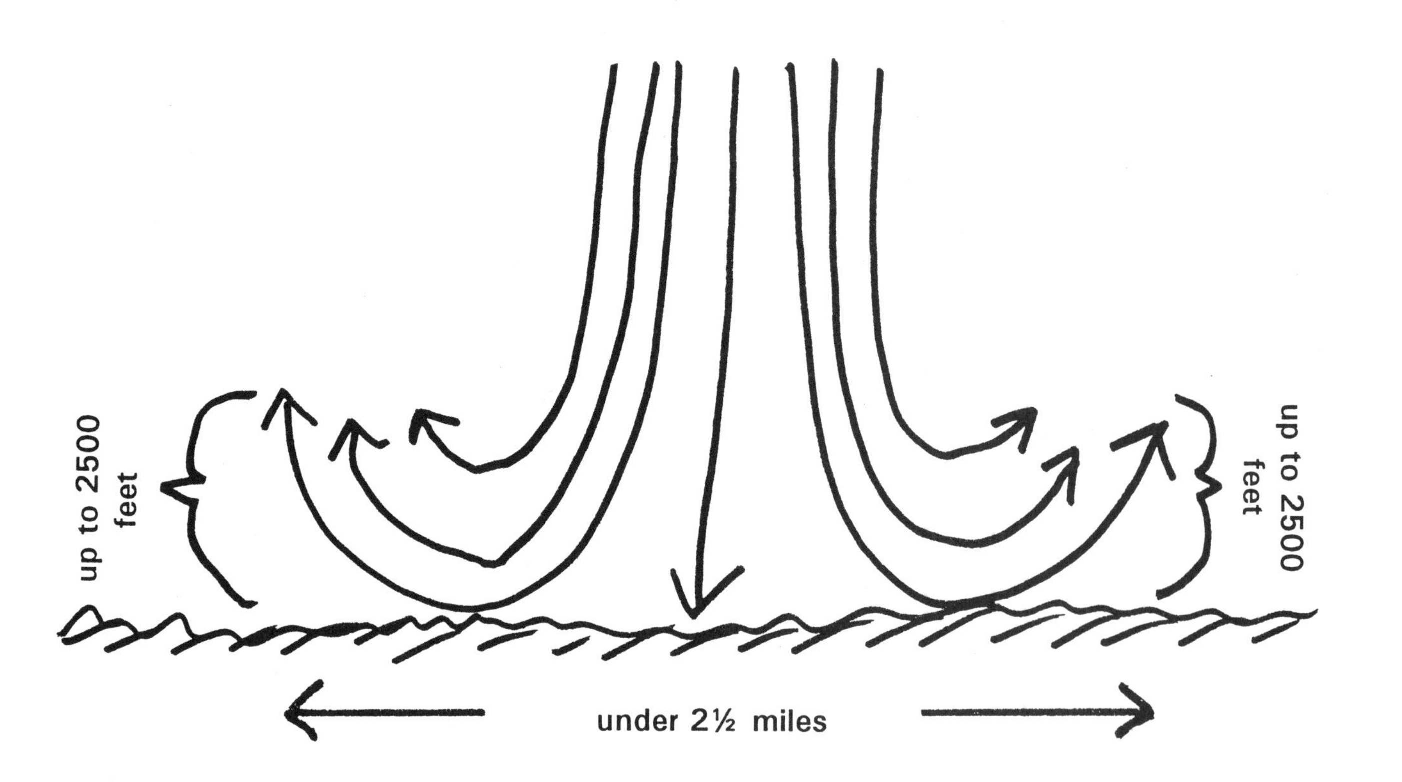

Figure 9-2. Note where the air is rising to the sides of the microburst's core. If houses are in this section, then the rising air can lift their roofs off similarly to how the rising air inside a tornado funnel lifts the roofs off houses. The air circulation of a tornado is both rotating and rising. The overhanging of a roof on the frame of the structure causes the roof to act like an airplane wing to get lifted up by winds that can exceed 100 miles per hour. This rising curl of air that develops after the microbursts splatters on the ground, can rise to as high as about 2000 to 2500 feet.

Nocturnal Thunderstorm Heat-Bursts

This strange weather phenomenon tends to occur in the Plains states, essentially between the Rocky Mountains and the Appalachian Mountains.

With the exception of some rolling hills and local low-elevation mountains such as the Ozarks in southern Missouri and northern Arkansas, this area is relatively flat. In the western portion, the land slopes upward as we approach the foothills of the Rocky Mountains.

Often, during the warmer half of the year, because of the flatness, what is called a "low-level jet" develops and lowers at night towards the surface. A low-level jet is a band of strong winds typically some 5000 feet above the surface. At night, the sun's effect of stirring the air up and mixing it is ended, and the "mixing layer", which is up to 2000 feet deep from the surface up, during the day, shrinks to being only a few hundred feet up. If a low-level jet is underway, then at night it will lower closer to the ground. The effect is to bring the stronger winds lower. If these winds are coming in from a southerly component, they will bring in warm, moist air which may include moisture evaporated into it from the Gulf of Mexico. The effect is to "destabilize" the area, which meteorologically means, in this case, the additional low-level heat and moisture primes the atmosphere for increased thunderstorm potential. How the atmosphere does this is explained later in the book. The "bottom line" is this: this destabilization process creates nighttime thunderstorms in the Plains states.

Thunderstorms develop in the afternoon and evening, but it is not unusual to have thunderstorms occasionally develop at night, especially in the Plains. What is unusual is the rare event that occurs in this nocturnal storms: a nocturnal heat burst.

Seemingly all of a sudden, in the midst of a raging thunderstorm with heavy rain, strong and gusty winds and possibly some hail, the precipitation stops and the temperature surges to well over 100 degrees. This is typically around 2 to 3 o'clock in the morning!

It may be 65 or 70 degrees during the rain, yet suddenly the temperature shoots up to 105 or 110 or even higher, stays there for perhaps 10 minutes, and then gradually falls back to where it should be.

People, cattle and other creatures caught outside during the rain and drenched by it are almost immediately dried. In the worst of these nocturnal heat bursts, the temperature has soared to over 130°F, scorching crops and threatening the lives of livestock.

Reports of these events occasionally surfaced every summer, but none was ever actually recorded as proof until June 20th, 1989 when it occurred at an official weather observing station at Pierre, South Dakota.

Below are data from the weather observations.

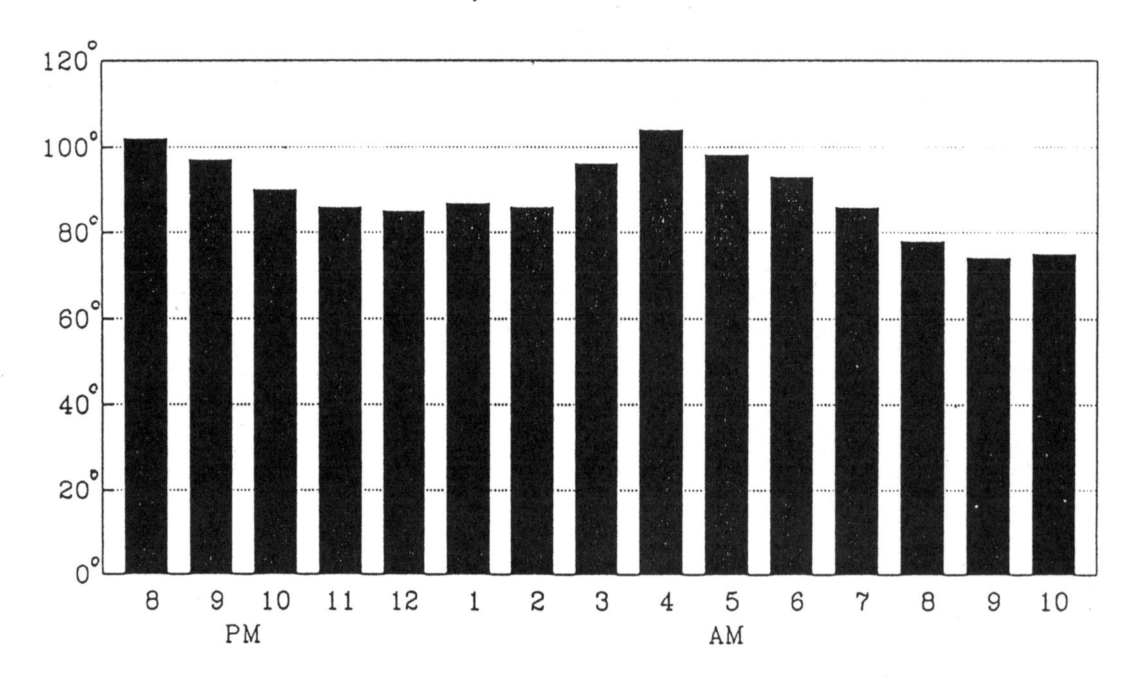

Figure 9-3. The overnight temperatures at Pierre, South Dakota during the nocturnal heat burst of June 20th, 1989. Times are in Central Daylight Time.

Note how the temperature surged from 86 at 1 a.m. Central Standard Time/2 a.m. Central Daylight Time, to 96° an hour later, and then to 104° in the middle of the night! There were thunderstorms occurring, with the peak wind gust during the heat burst of 61 mph. By after daybreak the temperature had fallen to a more comfortable 74°.

Examining the temperature sounding (the graph of the temperature in the vertical, taken by a weather sensor called radiosonde which is sent up by weather balloon), meteorologists noticed that there was a surface-based temperature inversion in place underneath the developing thunderstorm. An inversion is when the temperature warms up with increasing altitude. Near the surface, the norm is for the temperature to decrease with increasing height.

Thunderstorms contain rapidly rising air currents called **updrafts** and rapidly descending air currents called **downdrafts**. What appears to have happened to generate this thunderstorm heat burst is that some of the warm inversion air was transported up by updrafts towards the top of the thunderstorm...was which over 45,000 feet high...and then brought down by downdrafts. As air descends, it warms up. By the time this air reached the ground, it had warmed to 104°...at 3 o'clock in the morning (4 a.m. on daylight time)!

One of the worst cases yet reported of a nocturnal thunderstorm heat burst occurred in Texas in the 1940s, before the days of air conditioning. A nocturnal thunderstorm produced a heat burst in the middle of the night that caused the temperature to soar to an estimated 140°F. This would have been a new world's record, greater than the 136° recorded in Libya, if thermometers were there that recorded temperatures that high.

People woke up gasping, crops were scorched and cattle perished in the stifling heat. These thunderstorm downdrafts are air sinking through up to 8 miles, sometimes more, of atmosphere, warming and drying during the sinking process. By the time the downdraft smashes into the surface, it may be in excess of 130°F, according to some of the reports of the most extreme examples of this phenomenon.

A nocturnal thunderstorm heat burst similar to the Texas event described above is documented in the journal, Monthly Weather Review, for July 1909. This heat burst also occurred in Texas, resulting in crops burnt to a crisp and apples baking on trees.

As far north as Fargo, North Dakota, an airline crew on take-off during a heat burst reported 104° just above the surface shortly after sunrise on July 19th, 1977.

Thus, we are aware of this phenomenon, but we are a long way from daring to make a weather forecast such as this: "Partly cloudy tonight with scattered thunderstorms. Overnight low temperature about 70 degrees, but rising briefly to about 135 degrees around 3 o'clock in the morning." Leave your apples out to bake.

A journal called the Monthly Weather Review reported the following in its July 1909 issue:

"Between one and three o'clock on the morning of July 11, a strange phenomenon occurred in Cherokee, Okla. and vicinity. The thermometer recorded a rise in temperature to 136°. The effect of the hot winds on vegetation indicated that the center of greatest heat was a few miles south of Cherokee."

The following are some documented cases of these heat bursts from old historical files.

A probable rare case east of the Appalachians occurred in Boston, Massachusetts on September 23rd, 1815. It was reported as "a suffocating current of air".

From Lawrence, Kansas on September 23rd, 1882: "Fierce blasts burned the foliage of trees so that they crumbled to powder at the touch of the hand."

From Brownsville, Nebraska on July 29th, 1887: "Winds hot as the breath of a furnace; corn and vegetation burned up, apples baked on the trees in some instances."

Many of these reports are thanks to local newspaper documentations.

In Downs, Kansas on June 27th, 1890: "Hot winds burned the blades of corn, turned them white."

In Sulphur, Texas on May 30th, 1892: "Plants were scorched black."

On July 18th, 1896 a Dr. Cline reported a temperature of 140°F near Abilene, Texas in a heat burst episode. The 136° event at Cherokee, OK also had evidence of even higher temperatures nearby.

Nocturnal heat bursts are not just a phenomenon of the lower 48 states. One of thee events occurred at Killarney, Manitoba, Canada on June 30th, 1921.

Perhaps the worst case reported in the 20th century occurred shortly after midnight on June 15th, 1960 at Lake Whitney, Texas. The temperature shot up from the 70s to about 140°. Lightning was in the vicinity. The temperature shot up as the winds surged to about 100 miles per hour, causing considerable damage. The extreme heat caused car radiators to boil over. Cotton fields were pulverized. An unofficial thermometer reading at Kopperl, Texas reported at least 140°. Fortunately, this event was documented by a television station at Fort Forth.

Thus, we have sufficient evidence to state that nocturnal heat bursts exist, that they occur typically in thunderstorms, especially in decaying thunderstorms, which is when the storms are primarily downdrafts, that they occur mostly in the Plains states, and that they can cause the temperature to surge towards 140°F, and possibly even higher.

Chapter 10. WINDS IN A THUNDERSTORM

Others chapters discuss the various winds of a thunderstorm; however, here is an overview of these winds.

In an environment for severe thunderstorms, the low-level winds range from light to strong and gusty, with mid- and upper-tropospheric winds typically being stronger. A change of wind direction and/or wind speed with height (these changes are called wind-shear) is common in the environment for the development of severe thunderstorms.

For routine (non-severe) thunderstorms, the following conditions are typically observed during the storm's life-cycle:

CUMULUS-STAGE WINDS:
The local environment has typical surface wind speeds under 20 miles per hour. Updraft speeds in the cumulus cloud may reach 20 to 35 mph (8 to 15 m/sec).

MATURE-STAGE WINDS:
A gust-front or down-and-outrush of winds along the leading edge of the thunderstorm may exceed 30 mph. In severe thunderstorms, the gust front winds can exceed 50 mph, and when any winds at the surface in the storm are at or above 50 knots (58 mph), the storm is termed "severe" in the United States. In a severe thunderstorm, updrafts can exceed 50 meters/second or about 115 miles/hour, and downdrafts can be in excess of 50 miles per hour. In a routine non-severe thunderstorm, the updrafts in the mature stage are typically about 35 to 60 mph (15 to 25 m/sec).

DISSIPATING-STAGE WINDS:
Downdrafts take over as the storm dies, with typical speeds of 10 to 45 mph (about 5 to 20 m/sec).

THUNDERSTORMS PRODUCING HAIL:
The larger the hail, the stronger the updrafts need to be to sustain the hailstones in the cloud before they fall to the surface. Research has suggested that for hail from one-half inch to three-quarter inch diameter, the updrafts needed area bout 25 to 40 miles per hour; for hailstones of about an inch (2½ centimeters) in diameter, the updrafts need to be about 50 mph; and for hail of 3 inches in diameter, or softball- to grapefruit-size, the updrafts need to be at least 100 miles per hour, and can exceed 125 mph.

DOWNBURST AND MICROBURST:
Downbursts and smaller downbursts called microbursts, of wind, exceed 50 miles per hour. Recall that a downburst is a concentrated downdraft, and that a microburst is a concentrated downburst. Microburst winds are typically stronger than in the larger downburst. Microburst winds are typically from 60 to 100 miles per hour, but extremes have exceeded 100 miles per hour (160 kilometers per hour).

DERECHO:
This fast-moving type of thunderstorm produces a long swath of damaging winds which can reach or even exceed 100 miles per hour.

TORNADO:
The winds inside a tornado range from 40 miles per hour in the faintest of funnels, to over 300 miles per hour (over 480 kilometers per hour) in the most intense tornadoes.

WATERSPOUT:
A funnel forming over a body of water is not the same as a tornado that subsequently moves over a body of water. The tornado may still maintain its intensity for a while. A true waterspout, however, typically has winds of some 35 to 70 miles per hour, sometimes somewhat higher.

Chapter 11. THUNDERSTORM OUTFLOW BOUNDARIES

The outflow from the clouds to the surface from an approaching thunderstorm is called the thunderstorm's outflow boundary of the thunderstorm.. This rush of air is induced by the rainfall which drags down some surrounding air, creating the downdraft.

The outflow boundary is usually noticeable, since the surface winds suddenly increse to 25 to 40 mph (40 to 65 km/hr), although on occasion this rush of air can be severe, with winds in excess of 60 mph (95 km). Sometimes the outflow can spiral and form what has been termed a "gustnado" of strong winds, which is not a tornado. It is almost unheard of for a tornado to be found at the leading edge of a thunderstorm; tornadoes typically occur in the southwest portion of a thunderstorm in the Northern Hemisphere.

The outflow boundary is strong when the leading edge of the thunderstorm contains a low-level cloud that rolls and "boils" along, which is called a roll-cloud. Thus, when the leading edge of an approaching thunderstorm appears turbulent, it probably is, and the observer can expect a rush of strong and sometimes damaging winds.

Since colliding boundaries force air upwards, then when the outflow boundaries of two thunderstorms intersect, or when the outflow boundary of one thunderstorm intersects another boundary such as a cold front or sea-breeze front, then the result is rapidly upward vertical motion, especially if the collisions are occurring in moist unstable air. It is common for a new and rapidly-growing thunderstorm to form where intersections of such boundaries occur. Moreover, the rapid growth and strong updrafts tend to make these areas of merging storms become severe thunderstorms.

Even when a thunderstorm is dissipating, its outflow may continue for a while, and if moving into moist, unstable air, may initiate further convection.

Thus, thunderstorm outflow boundaries are important mesoscale weather features.

How strong the outflow's winds are depends on the strength of the downdraft caused by the precipitation drag on air parcels as the precipitation falls, and on the cooling that occurs when latent heat, produced by the condensation process when precipitation is formed, is absorbed by the air. This cooling process makes some of the environment in the thunderstorm cooler, which destabilizes it further, thus allowing rising parcels nearby to keep rising since they will be warmer than the environment they are rising through. This sounds somewhat confusing, but the "bottom line" is that the conditions of the local environment determine how strong is the outflow.

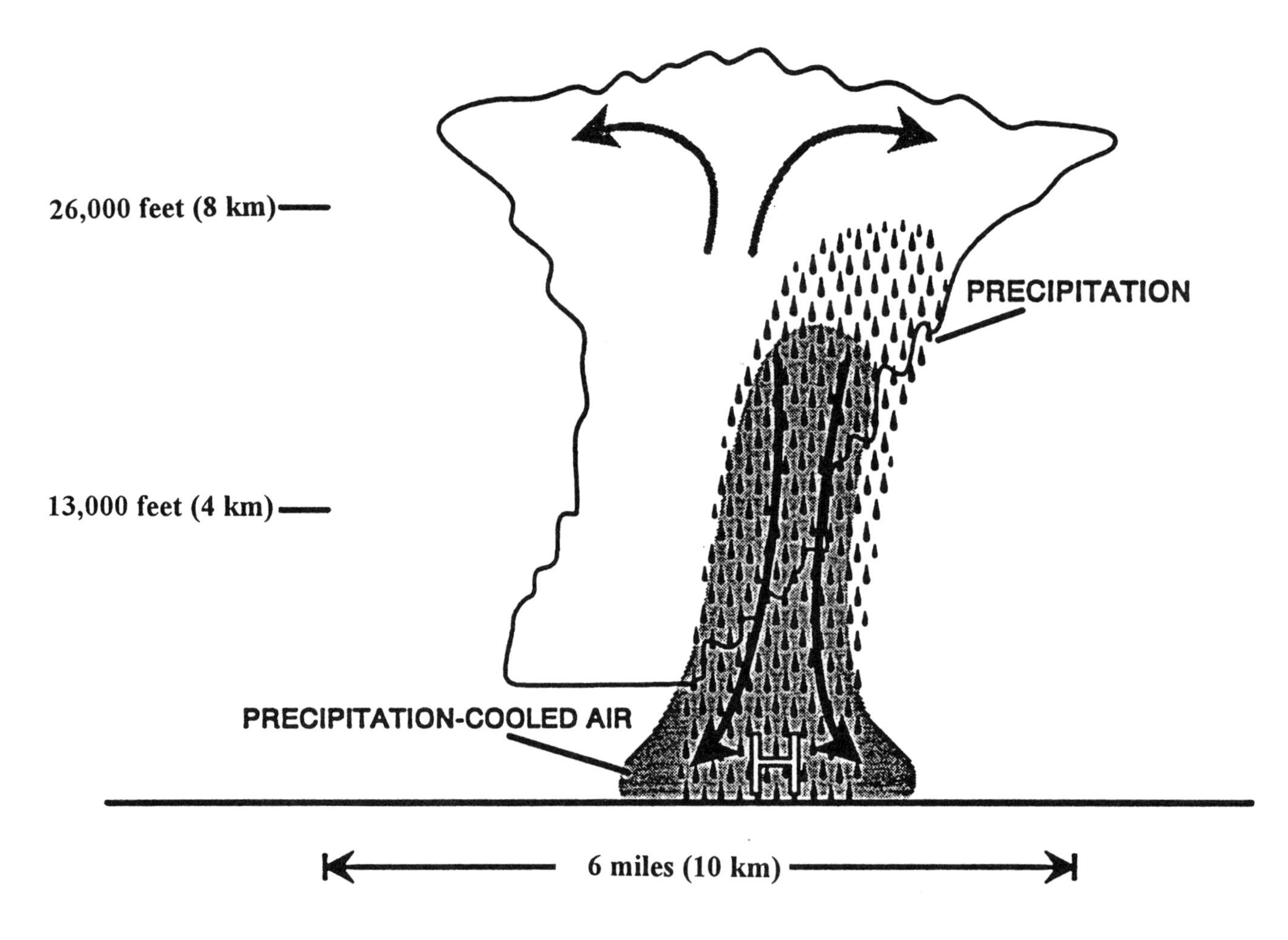

Figure 11-1. The outflow boundary or gust-front is at the surface along the leading edge of the thunderstorm. At the surface, under the downdraft that causes the outflow, is a weak small-scale (mesoscale) high pressure system caused by this descent of cooler air. At the surface under the updraft region is a mesoscale low pressure system caused by the lowering of pressure as the air rises from the area. Thus, there is a mesolow and a mesohigh, in terms of air pressure centers, associated with a typical thunderstorm cell. (source: NOAA)

Thus, in the thunderstorm, the downdrafts bring to the surface colder air, just as updrafts are rising warmer air. Therefore, **the leading edge of a thunderstorm outflow boundary is analogous to a mini-cold front, and is referred to as the thunderstorm gust-front.**

On a typical summer afternoon in much of the lower 48 states, it is common, during convective episodes, to observe the temperature plummet rapidly from the 90's (F.) To the 70's or 60's as the thunderstorm gust-front passes through. The wind shifts and the pressure rises after the outflow boundary has passed, just as if a mini-cold front has passed through. After the thunderstorm passes, the effects of the passage of the gust-front and the rest of the outflow disappear.

The outflow boundary is usually associated with a rainy downdraft; the precipitation begins just after its passage.

Even when the thunderstorm itself is in the dissipating stage, its outflow boundary may continue moving along, well ahead of the dying rain area. As stated before, if it intersects another boundary in moist unstable air, then new convection may be initiated.

During the thunderstorm, the outflow boundary spreads out in all directions, but mainly in the direction of movement of the storm. Thus, some areas may experience the gust-front passage but not receive precipitation, depending on where they are in relation to the storm and its movement.

The descending cold air in the outflow is a lifting mechanism for the warmer air it pushes into. Moreover, the outflow boundaries by themselves are typically zones of converging air; thus, new convection may develop along them anyway.

Since most convective episodes have more than just one thunderstorm cell, there may be several surges of gust-fronts. Moreover, the individual cells with their own respective outflow boundaries may ultimately merge and produce a "bubble high" (small high pressure area) thunderstorm complex. A bubble high is a mesoscale area of higher pressures and lower temperatures than its surrounding area. The combined gust-fronts form the leading edge of the bubble high thunderstorm complex. This leading edge is frequently analyzed as a squall line. Sometimes a cold front is initiating a squall line of thunderstorms and the cooler air aloft advances faster and farther than the surface front, which destabilizes some of the area ahead of the front (with the cooler air moving in aloft), and causes a pre-frontal squall line to also form, perhaps some 50 to 100 to even 150 miles ahead of the squall line along the cold front.

The pictures on the next page show examples of cloud forms that occur with strong gust-fronts. A gust-front is the leading edge of an outflow boundary .

Figure 11-2. A roll-cloud, which is usually detached from the main cloud and is tube-like, is near the bottom of this photograph, and occurs with the gust-front which is the leading edge of the thunderstorm outflow boundary. Sometimes this roll cloud rotates slowly around a horizontal axis, which is the forward edge rising and the rear side sinking. Observers sometimes think this is a developing tornado, but tornadoes virtually never come out of the front advancing edge of a thunderstorm. (source: NOAA)

Figure 11-3. A "shelf-type" outflow boundary cloud, with this cloud attached to the base of the main cloud above it. (source: NOAA)

Chapter 12. SLANTWISE CONVECTION

Most convection, since it happens so rapidly, is either straight up or nearly vertical. The change in wind direction and speed with height causes some updrafts to have some tilt.

From our observations of convective development, we know that in mid-latitudes during the warmer half of the year, showers become thunderstorms when the cloud tops grow past the level of about 25,000 feet (about 8 kilometers) high, assuming that the bases of the clouds are from about one to four thousand feet high. For higher based thunderstorms, such as the mid-level bases of 8 thousand to 14 thousand feet up found sometimes in semi-arid and arid areas such as the U.S. western high plains and over the U.S. desert southwest, the depth of cloud starting from the level of the cloud base, still needs to be about 25,000 feet thick, thus giving cloud tops above about 35,000 feet before the showers produce lightning and thunder and are therefore now thunderstorms rather than showers.

Even though most convection ranges from vertical to slightly tilted, the rest of convection, therefore, occurs at a slant, and is thus called SLANTWISE CONVECTION.

In slantwise convective episodes, consider the usual case, as with vertical or nearly-vertical convection, of cloud bases within a few thousand feet of the surface, and the updrafts resulting in cloud growth that changes the showers into thunderstorms, yet the cloud tops have not passed the threshold of about 25,000 feet high, and may in fact only be 10 thousand to 18 thousand feet high. In these cases, the updraft of each cell is tilted, but the range of the updraft is still at least 25,000 feet. When you measure the height of the cloud mass vertically, it may be, for example, only 15,000 feet high, but when you measure the depth of the slanted updraft, it may be 25,000 feet long, thereby producing lightning and thunder with the precipitation.

Most slantwise convection events occur during the colder half of the year and often are imbedded within the cloud masses of major low pressure systems that cause widespread rain and snow storms. The special case of thundersnow is discussed in the following chapter.

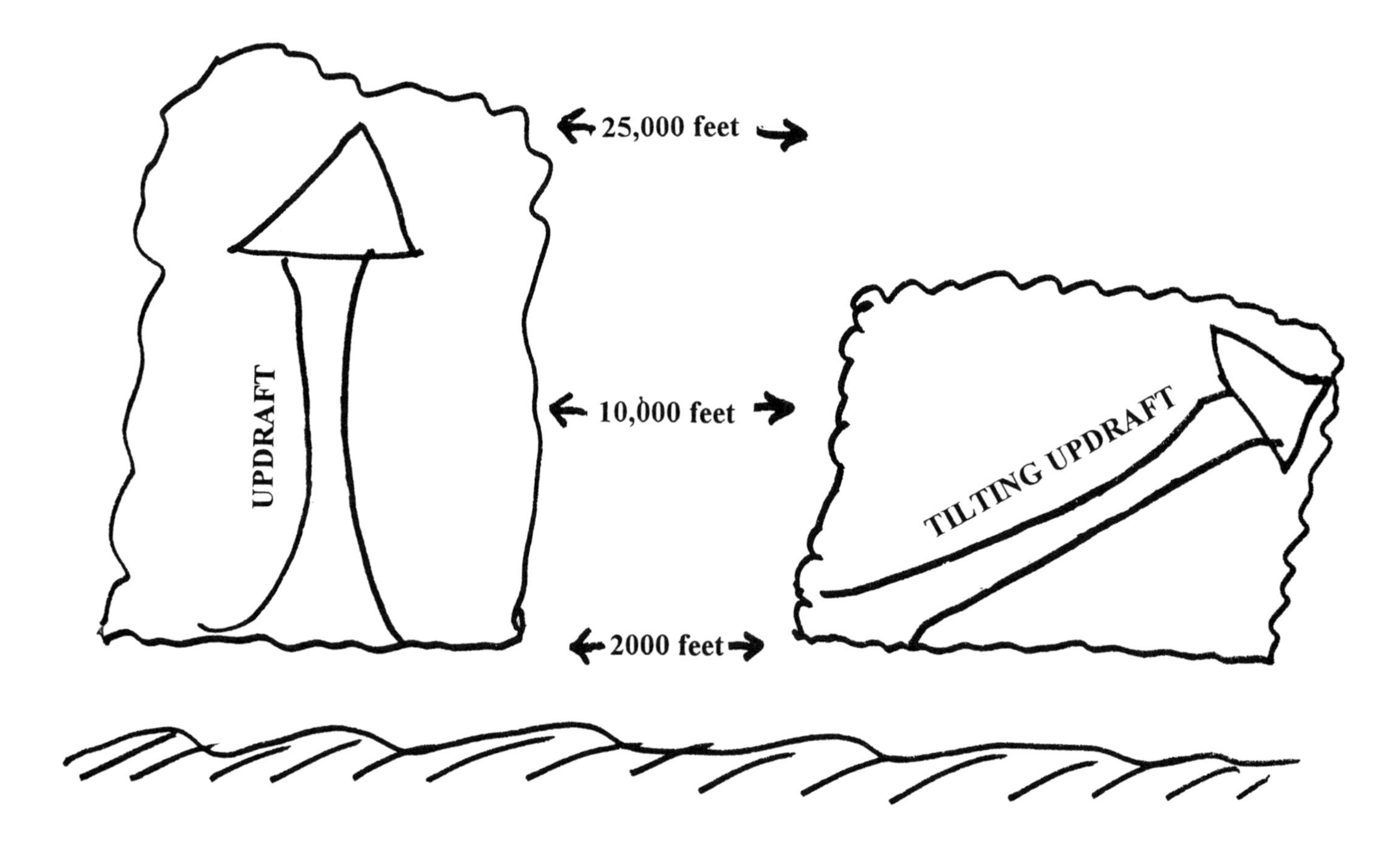

Figure 12-1. Vertical or nearly-vertical convection compared to slantwise convection. In middle-latitudes, the depth of the convective updraft must be about 25,000 feet long, either vertical or at an angle, for showers to evolve into thunderstorms. The first lightning and thunder typically appear when the fetch of convective cloud growth passes the threshold of about 25,000 feet (about 8 kilometers).

Chapter 13. THUNDERSNOW; LAKE-EFFECT THUNDERSNOWSQUALLS

Figure 13-1. Some of the greatest rates of snowfall occur during episodes of thundersnow. This is because this snow is falling from convective clouds, which have a much greater vertical extent than stratiform clouds of a well-developed low-pressure system. Snowfall rates of two to five inches per hour are common in thunder-snow events, with eleven inches falling in one hour in Oswego, New York from a thundersnow episode. (source: NOAA)

To have convection, the local environment needs a lifting mechanism for the air parcels, moisture and instability, with no cap aloft, as discussed earlier. However, when the convection is slantwise, as long as the convective cloud fetch is at least some 25,000 feet, even if the cloud tops do not exceed say 12,000 to 15,000 feet, then there can be a cap aloft vertically, since the convective growth would still be sufficient for thunderstorm generation. Moreover, there are occasions when intense lift forces thundersnow even when the local environment is not quire unstable. We have seen this happen, for example, when diverging-air quadrants of jet-streaks of the subtropical and polar jet streams overlap each other to greatly enhance the lifting of parcels and layers of air. (Note: The book, "WEATHER MAPS - How to Read and Interpret all the Basic Weather Charts", contains a detailed explanation of the role of the jet-stream in vertical motions. This book is available from Chaston Scientific, Inc. at the address given on page 1.)

The "bottom-line" is this: when a winter storm, or when the winter lake-effect snow mechanism, also contains thunderstorms which are producing snow, then that area will receive more snowfall than it otherwise would.

When thunderstorms are imbedded within a major winter cyclone, or within a non-major "overrunning" situation (warm-air in a widespread overrunning or lifting over a dome of cold high pressure), and they were not anticipated, then the snowfall amount for that area is likely to be underforecasted. How long thundersnow persists over an area is the key; two hours of it can yield, for example, several more inches of snow than would have otherwise occurred.

From weather satellite images, meteorologists can spot areas where thundersnow may be developing. In visible pictures, areas where the tops of clouds are overshooting the massive cloud area of the low pressure system are likely embedded convection. This may not necessarily be slantwise convection, but overshooting tops likely indicate stronger vertical motion compared with the vertical motions in the rest of the cloud mass. In the enhanced infrared satellite pictures, which show the temperatures of the cloud tops, meteorologists look for small areas of colder tops, meaning these are overshooting tops.

Thundersnow often occurs as a narrow swatch of very heavy snow as the imbedded thunderstorms move along with the motion of the low pressure system. At about 10,000 feet above the surface, a strong synoptic-scale vertical motion within the stratiform steady rain or snow clouds is typically on the order of one-fifth of a mile per hour (about 10 to 20 centimeters per second), but convective updrafts for thundersnow can be in excess of 35 miles per hour (about 55 kilometers per hour). Thus, the convective-scale lifting is some two orders of magnitude greater than the synoptic-scale lifting.

These convective regions have far greater convergence of air in low- and mid-tropospheric levels than in the surrounding cloud mass, and therefore accumulate greater amounts of moisture that is transported in by these converging winds. The result is very heavy precipitation in the convective areas. Thus, being able to forecast thundersnow is critical since the snowfall rates in such episodes high and can be very great. When significant thundersnowfall persists for over an hour over the same region, snow piles up very rapidly.

LAKE-EFFECT THUNDERSNOWSQUALLS:

The snowfall that occurs from "lake-effect snow" episodes can also be high, but can be especially high when some of the lake-effect grows into thundersnowsqualls. A snowsquall is a burst of exceptionally heavy snowfall.

Firstly, here is an explanation of lake-effect snow:

When very cold arctic air passes over the relatively warm Great Lakes of the United States, especially from late autumn into early winter when the lakes are free of widespread icing over, clouds form over the warmer lakes and these clouds tend to produce copious amounts of snow. When the dewpoint of the cold air is at least about 18 degrees colder than the surface temperature of the Great Lake it is passing over, the evaporation of moisture from the lake condenses into clouds, and these clouds have vertical growth because of the unstable conditions (warm at the bottom and much colder as we rise up for several thousand feet). This situation is analogous to what you see when you fill a bath tub to the top with hot water. There is evaporation, and the moist air condenses in the cooler air over the tub to form a fog (a cloud).

When the low-level winds steer the lake effect snowclouds inland, the heavy lake effect snows dump their snow not just over the lake but also onshore. The heaviest lake effect snowfall typically occurs about 10 to 20 miles inland from the coast. The land is higher in elevation than the lake's surface, so since the air is moving onshore already saturated, it must rise up and over the land and its hills, etc. Any additional lift to the saturated air squeezes out (condenses out) even more moisture to increase the snowfall rate. This phenomenon is known as frictional convergence. With frictional convergence, air parcels pile up and ascend higher terrain, condensing out more snow. Thus, the heaviest lake effect snows (which can be three or four FEET and sometimes more, within several hours!) typically occur not along the shoreline but some 10 to 20 miles inland.

Ironically, some of the heaviest snows in the lake-effect belt of the Great Lakes occur on the back of the low pressure system as the storm is pulling away, since behind it is the influx of very cold air from a very cold high pressure system moving in behind the low.

In fact, very cold air passing over any large and significantly warmer water body can produce lake-enhanced or ocean-enhanced snows. For example, when the upper-air pattern causes a winter blast of arctic air to spill over the western mountains and make all the way over to Puget Sound in Washington state, then ocean-enhanced snowfalls can occur, similar to the lake effect snows of the Great Lakes. Actually, any substantial water body is such a situation can produce or enhance snowfall.

For example, on occasion lake effect snowstorms will be generated by very cold air passing over the Great Salt Lake in Utah. The low-level wind direction determines where on land the heavy snow falls.

Some extremes of lake effect snowfall include over 100 inches during one episode on the Tug Hill Plateau southeast of the eastern end of Lake Ontario, in New York State, nearly six feet at Chardon, Ohio from a November 1996 Lake Erie lake effect storm, and the famous Buffalo, New York lake effect blizzard, accompanied by high winds, in January 1977, in which snow drifts covered telephone poles and even entire houses. After the storm, people had to dig tunnels through the snow from their front doors to the streets, which were eventually plowed, just to get out of their houses! The upper peninsula of Michigan is a prime snow-lover's area, due to frequent lake effect snows from Lake Superior; it is common there to have snow seasons with over 200 inches of snow for the season. Thus, lake effect snows can be quite impressive. Now, add thundersnowsqualls and the amounts are even more impressive.

For example, one thundersnowsquall at Oswego, New York, on the southeast tip of Lake Ontario, dumped eleven inches of snow in one hour, as part of a 57-inch snowstorm that fell in about six hours. Visibility was zero from falling and blowing snow, and flashes of lightning accompanied by muffled thunder resulted in an eerie yet beautiful event.

The cloud tops of lake-effect snow are often below 5000 to 8000 feet, so that a weather radar even only sixty miles away will have its radar beam overshooting the clouds top from that distance, such as Buffalo New York's radar beam overshooting the tops of such an event over Rochester, New York, just 60 miles away, even though it may be snowing at three to five inches per hour at Rochester! Thundersnowsqualls, even if through some slantwise convection, would have their higher levels intercepted by the radar beam. Often, in thundersnowsqualls, cloud tops are only 12,000 to 16,000 feet high, which means that the actualy convection is on a slant, yielding at least 25,000 feet of convective development (convective fetch).

Chapter 14. THUNDERSTORM CLASSIFICATION

Most thunderstorms are "ordinary" or "routine" non-severe storms, most of which form in warm- to- hot and moist air masses, and are also called "air-mass thunderstorms". They need a "trigger" to initiate the lifting and must be in a locally unstable environment with no mid-tropospheric cap. All other types of thunderstorms also have unique characteristics, which therefore allows us to develop a classification for various types within the thunderstorm family. Also, air mass or ordinary thunderstorms can also evolve into any of these other types of thunderstorms. An important point to keep in mind is this: *"what determines the potential severity of a thunderstorm is the strength of the updraft; in general, the stronger the updraft, the stronger and more severe is the thunderstorm likely to be."*

a. ordinary thunderstorm (air-mass thunderstorm)

Much of the discussion in this book up to now is of the most common thunderstorm, air-mass or ordinary (routine) thunderstorms. The ordinary air-mass non-severe thunderstorm contains heavy rain, and some contain small hail and gusty winds, but below the severe criteria. (Recall that a severe thunderstorm is a thunderstorm which produces wind gusts of at least 58 mph and/or hail of at least 3/4 inch diameter.)

Below is a vertical profile of the life-cycle of an ordinary thunderstorm.

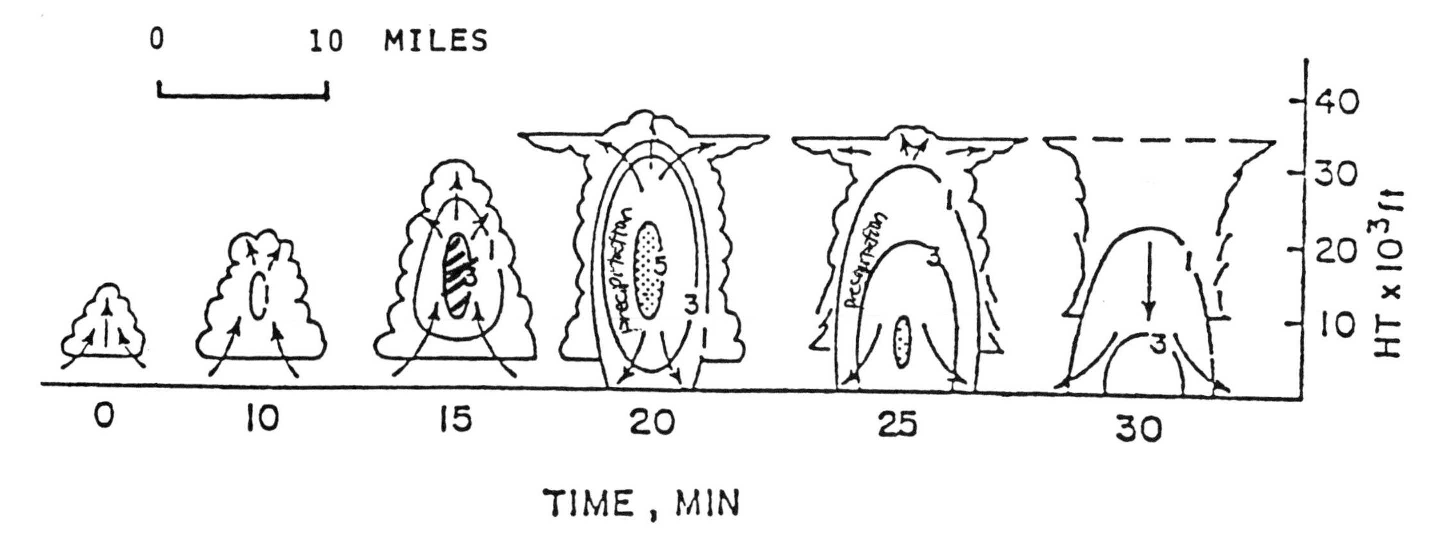

Figure 14-1. A time-lapse schematic of an ordinary air-mass thunderstorm. This data is compiled mainly from radar depictions of such thunderstorms. The darkest shaded-in area shows the region of heaviest precipitation.

b. pulse thunderstorm

A distinct type of thunderstorm than grows from nothing to mature more rapidly than does the ordinary thunderstorm, and is often briefly severe, and subsequently dies rapidly is the pulse thunderstorm, so-named because its life cycle is a rapid pulse of growth and decay. The entire storm may last as little as under thirty minutes from initiation to end, but can last for somewhat over an hour. When the pulse thunderstorm is severe, even the longest lasting one typically has its severe weather persisting for some 10 to 30 minutes.

The cause for this type of storm is an intense, concentrated updraft, which often has speeds in excess of 70 mph (over about 110 km/hr).

The first radar echo of precipitation typically appears at between about 20,000 and 30,000 feet above the ground, which is some 10,000 feet above the first precipitation echoes from ordinary thunderstorms.

The most intense precipitation core is also higher in the clouds in the pulse thunderstorm and maintains continuity as it descends to the ground. This core often has hail at the surface, which sometimes exceeds two inches (5 centimeters) in diameter, even though it lasts briefly.

Radar observations of such storms indicate that when this most intense core of precipitation extends to or above 30,000 feet above the ground in the cloud, the storm is likely to be severe.

Since this type of quick up-and-down development and decay scenario is caused by an intense, concentrated updraft, the local environment must have no or little vertical wind shear in direction and speed.

Aloft, inside the storm, there is typically a high liquid water concentration and also, often, large hail. Turbulence aloft can be extreme.

The reason that the precipitation first appears higher in a pulse thunderstorm than in an ordinary thunderstorm is because the updraft is stronger in the pulse storm, which carries the condensed moisture higher. We may surmise from this fact that ***we can identify the potential for any thunderstorm to become severe by whether its first radar echoes of precipitation are at a higher level than in the nonsevere thunderstorm.***

The pulse thunderstorm can generate damaging surface winds and occasionally weak tornadoes. Powerful tornadoes are rare, with such strong ones almost always coming out of a supercell thunderstorm, which is discussed later in this chapter.

Strong winds aloft play a key role in the generation of tornadoes, and the lack of such winds aloft in a pulse thunderstorm is a key reason why pulse thunderstorms rarely produce tornadoes, and if they do, they are typically weak.

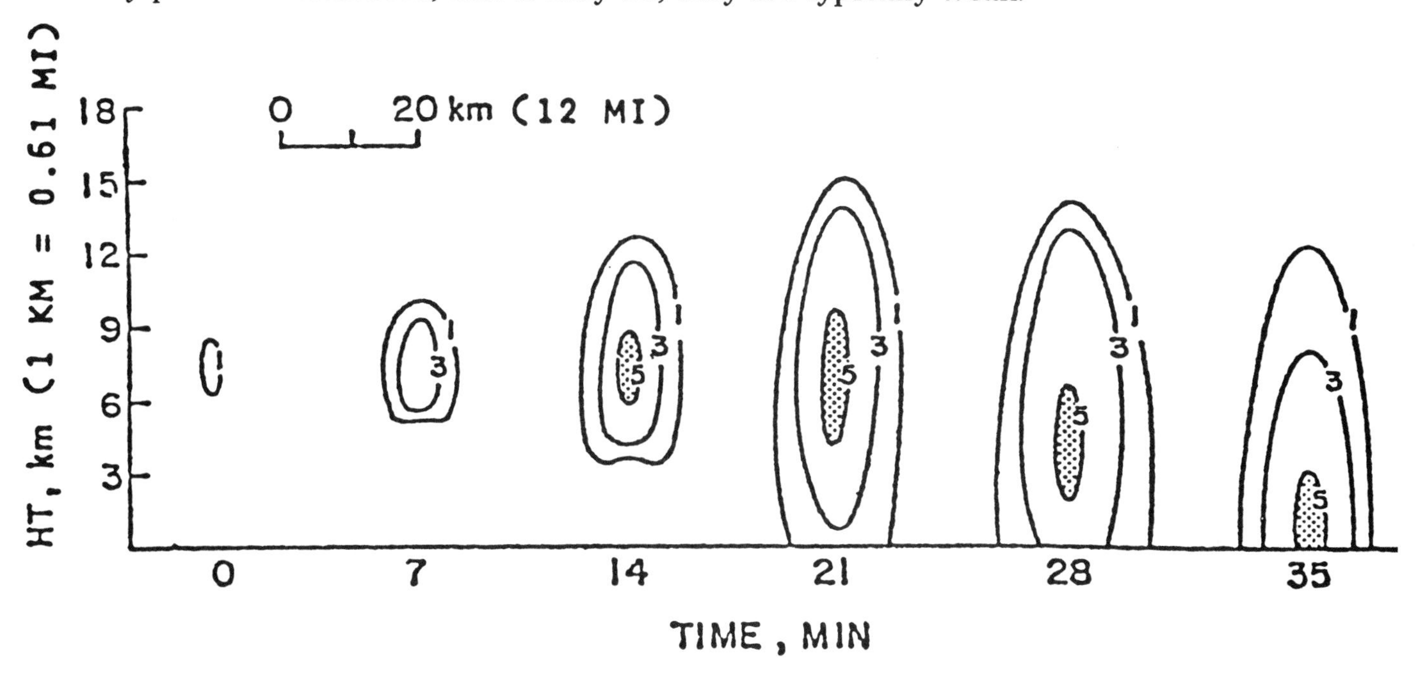

Figure 14-2. Precipitation aloft in a single cell pulse thunderstorm. The shaded-in area is the core of the strongest precipitation. This information is gathered from weather radars vertically scanning the depth of thunderstorms, and detecting where inside the cloud is the precipitation. The intense, concentrated updraft develops rapidly, generating intense precipitation, often briefly including large hail, at some 20,000 to 30,000 feet above the ground level, all of which descends to the ground. The pulse thunderstorm forms rapidly and decays rapidly in a quick up-and-down evolution, in a local environment of no or very little wind direction change in the local vertical environment and no or very little wind speed increase in the local vertical environment, in which the storm forms.

Operational meteorologists who are responsible for issuing severe weather warnings monitor the local environment through vertical sounding data of the loer atmosphere, and look for rapid changes of the local supply of water vapor for storm development, and for any changes in the wind field. These changes typically can have a significant impact on the type of thunderstorm that forms. We know, for example, that increasing the low level temperature and/or low-level moisture supply (dewpoints) makes that region more buoyant, thus enhancing the lifting of air parcels. The pulse storms form in areas of moderate to strong buoyancy.

Observations show that the pulse thunderstorm tends to move with the average wind from the surface to about 15,000 to 25,000 feet (5 to 8 kilometers) up.

You may experience an entire afternoon of pulse thunderstorms popping up and dying, with most of them producing briefs burst of severe weather, usually large hail.

c. multi-cell thunderstorm

The multi-cell type of thunderstorm system is often severe, and indeed is a common type of severe thunderstorm. It appears that ***the greatest percentage of severe thunderstorms are multi-cell thunderstorms, although the most severe type of thunderstorm is the supercell type***, which is discussed next.

The multi-cell storm contains a number of individual cells or a series of strong cells, each in its own stage of development. Thus, in an ongoing multi-cell thunderstorm system, some cells may be in the cumulus stage, others in their mature stage and the rest in the dissipating stage.

Strong updrafts cause some multi-cell storms to be severe, but merging cells are also an important factor. Recall that there is lift ahead of a thunderstorm outflow boundary; therefore, intersecting outflow boundaries greatly enhance lifting, and may indeed cause rapid new convection. Moreover, new cells form where the conditions are most favorable, such as where the air is most buoyant (unstable). A preferred area for new cell formation is at the right or right-rear flank of the storm system.

The more unstable the air, the more likely are multi-cell storms to become severe; also, when vertical wind shear also exists, the multi-cell systems are more likely to be severe.

Individual cells within the multi-cell cluster typically move in one direction, such as southwest to northeast, as the entire system itself moves in its direction, such as northwest to southeast. Therefore, the resultant motion of any individual cell is the combination of both motions. Moreover, new development and decay components are included in how the system appears in determining the resulting movement of the cluster itself.

A multi-cell thunderstorm system has a series of cumuloform towers which is called the flanking line. If a part of the cloudbase in a flanking line lowers, which is called a wall cloud, and subsequently rotates, then a tornado may be trying to form.

Weather satellite images will show any cloud tops that overshoot the main cloud mass. These overshooting tops are typically associated with the strongest updrafts.

As these updrafts push into the stable lower stratosphere, they cannot continue to support the overshooting tops, with the result being a subsiding of these tops, and sometimes a rapid subsiding or collapsing of same. It is possible that rapidly collapsing tops are associated with powerful downdrafts that can lead to microbursts.

In and near the core of the updraft, the precipitation is held aloft for a while. On a vertical scan of this region by a weather radar, the radar often depicts a weak echo region which eventually may be bounded or surrounded by precipitation on both sides and above it. A bounded weak echo region can be a precursor to possible tornado-genesis right there.

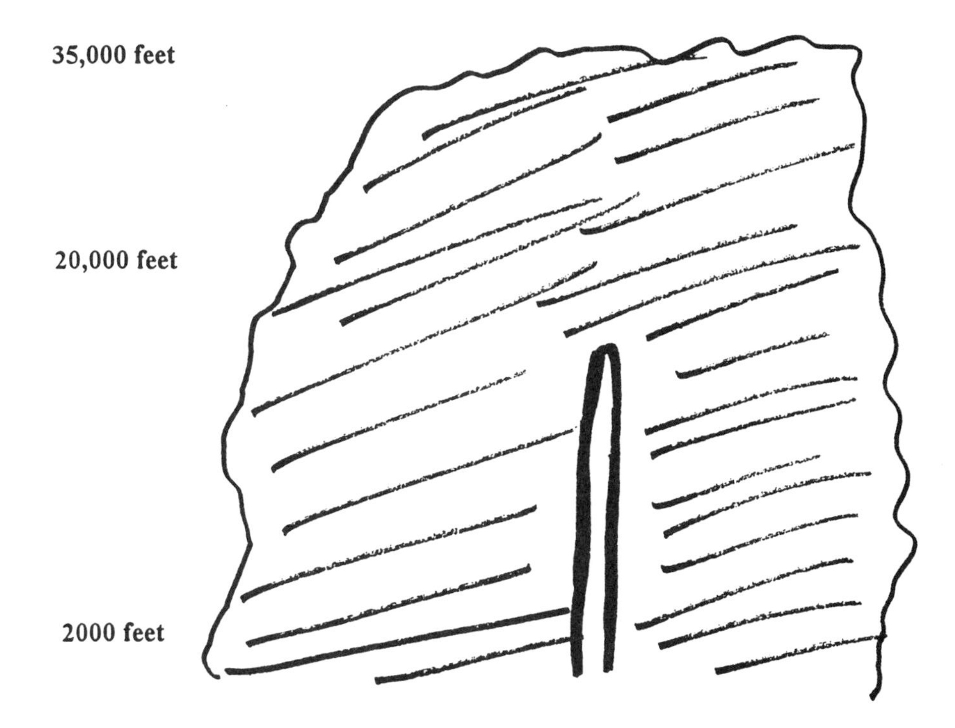

Figure 14-3. A bounded weak echo region. A weak radar echo means little or no precipitation is occurring there, which is due in this case to a powerful updraft that is temporarily holding the precipitation aloft. Sometimes a weak echo region within a thunderstorm is a precursor to tornado development, since it may indicate that air is spiraling into that region.

In a multi-cell thunderstorm system, new cells usually form at the upwind edge of the cluster, which in the Northern Hemisphere is most often the west to southwest edge of the cluster, the mature cells are typically in the middle of the cluster and the dissipating cells are at the downwind edge of the cluster, which in the Northern Hemisphere is most often the northeast to east edge of the cluster.

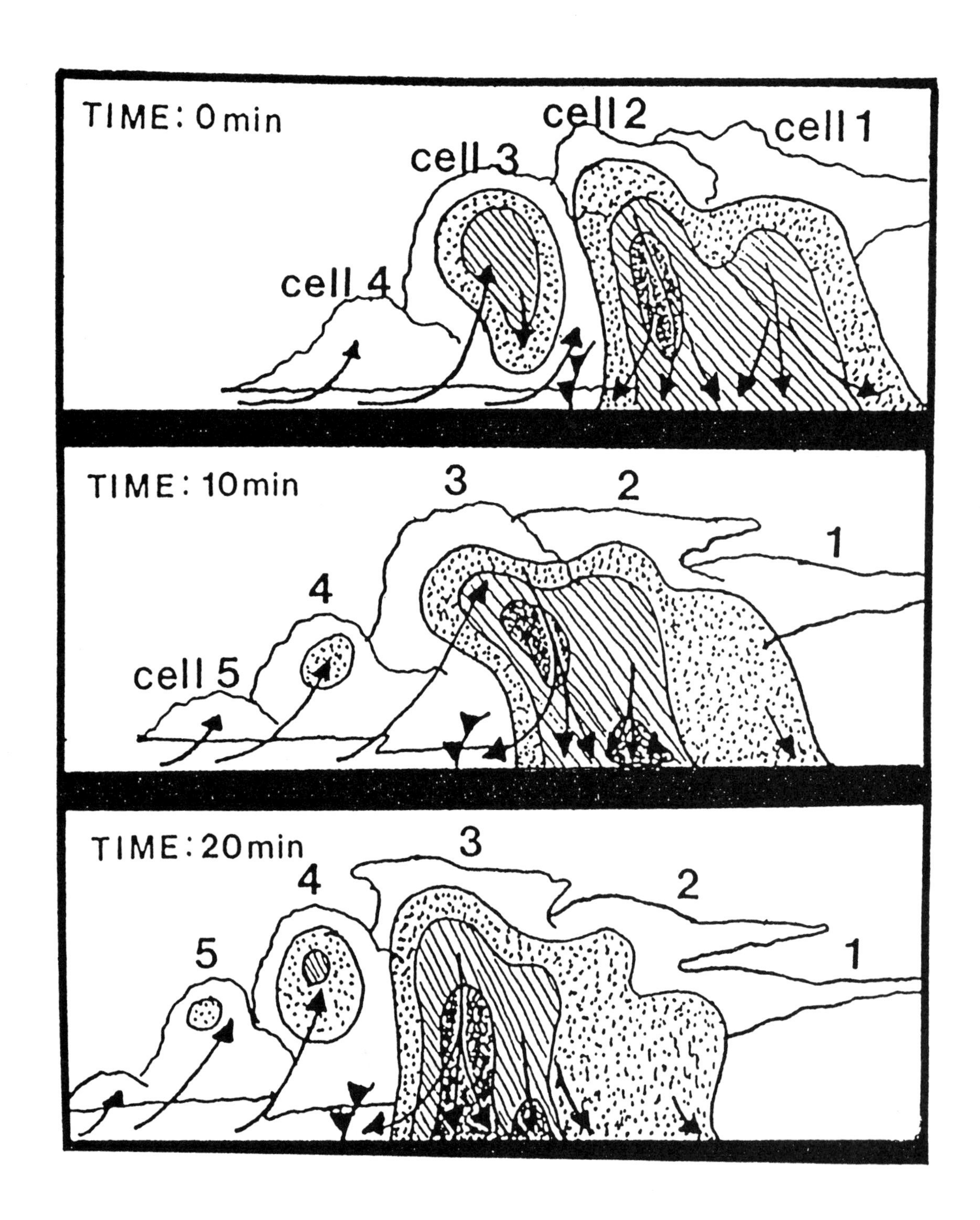

Figure 14-4. **The propagation of a multi-cell thunderstorm cluster.** The shaded areas are precipitation, with the darker the shading representing the heavier the precipitation.

Note how rapidly the cells change in just ten-minute intervals. (source: NOAA)

Each cell in the cluster has a typical life span of some 20 to 40 minutes, but the entire cluster usually persists for up to several hours.

In terms of severity ranking, the multi-cell thunderstorm is usually more intense than the ordinary thunderstorm, but not as severe as the supercell thunderstorm.

Multi-cell storms can produce heavy rainfall, especially if number of cells mature over the same area. They can produce hail up to about golfball size, can also produce downbursts and sometimes microbursts, with winds up to about 80 miles per hour, rarely higher, and also may produce weak tornadoes. Observations tend to indicate that *the most likely location of severe weather in a multi-cell thunderstorm is around the updraft/downdraft interface in the mature cells.*

The new cells in the system can develop every ten or so minutes in the right or right-rear flank of the system; this is the area where there is low-level inflow of air(new updrafts).

The actual resultant motion of the entire system is often somewhat to the right (in the Northern Hemisphere) of the mean wind flow from the surface to some 18,000 feet above the ground, because of new cells developing on the low-level inflow side and old cells dissipating on the other side.

d. supercell thunderstorm

A supercell thunderstorm is an intense long-lived thunderstorm that causes the most severe convective weather.

Supercells occur less often than the multi-cell thunderstorms, but they are the type of thunderstorm that is capable of producing the largest, most severe tornadoes. Supercells also can produce large to very large hail (hail of two inches [5 centimeters] or more in diameter) that can persist for many minutes, as well as smaller hail that can persist for over an hour. Supercells are also capable for producing wind gust in excess of 100 miles per hour (160 kilometers per hour).

Although individual air mass (ordinary) thunderstorms have a typical life-cycle of form some 20 to 60 minutes, a supercell has a typical life-cycle of about three hours, and they sometimes last as long as about six hours.

The sustained updrafts of supercells are from 50 to 100 mph or greater. Updrafts over 150 mph (240 km/hr) occur in the most intense supercells that produce hail of grapefruit or softball size.

Supercells tend to move and build into a more unstable environment.

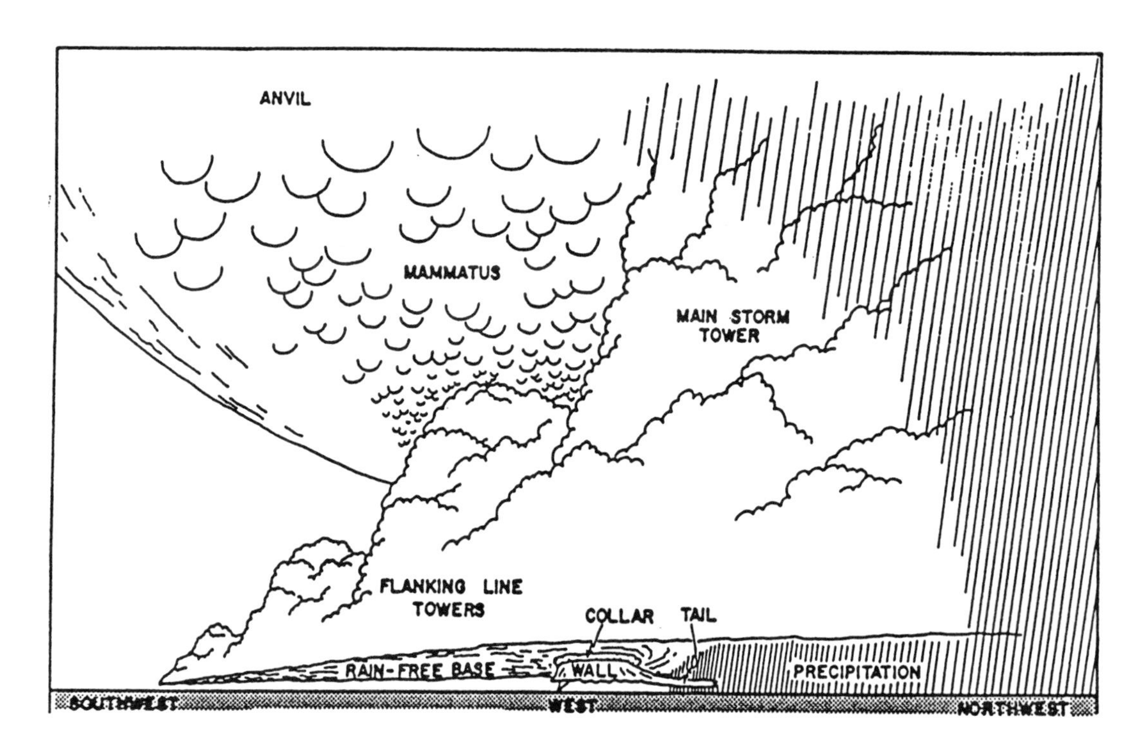

Figure 14-5. A sketch of a side-view looking northwest in a Northern Hemisphere supercell thunderstorm. This supercell is producing a tornado. (source: NOAA)

Most supercells move to the right of the 700-millibar (about 12,000 feet) steering winds, whereas most ordinary air mass thunderstorms move, as a first approximation, with the actual 700-millibar flow.

Moreover, supercells tend to move more slowly than the environmental winds.

Therefore, a supercell tends to move to the right of , and more slowly than, any neighboring ordinary thunderstorms.

Weather observations of environments in which supercells form, show that the vertical wind profile typically is one of the wind direction veering (turning clockwise) significantly as we ascend from the ground to the cloudbase. Thus, for example, the wind along the surface may be from the south, but some 2500 feet up if that is where the cloud bases are forming, the wind direction may be from the west-southwest.

Transformations of thunderstorm types occur also. Thus, a multi-cellular thunderstorm system may temporarily evolve into a supercell, or it may evolve into a supercell that persists for up to some six hours. Indeed, many multi-cellular thunderstorm systems evolve into supercell thunderstorms. Typically, one cell on the right flank of a multi-cell system becomes intense and severe. Moreover, supercells may evolve from other types of thunderstorms.

When a radar weak-echo region develops and becomes a bounded weak-echo region (as described earlier), then the hook echo on the radar precipitation display shows up, which is one radar signature of a likely tornado. When the bounded weak-echo region begins filling with precipitation echoes, then the updraft is weakening, and the pendanat swings cyclonically (counterclockwise in the Northern Hemisphere), merging with the parent precipitation echo. The cloud top collapses and the severe weather threat is then the greatest for a tornado, for the largest hail and for the strongest downburst or microburst.

An important characteristic of a supercell thunderstomr is that its updraft rotates. You can see the rotation in the cloud.

A supercell develops in very unstable air, usually with a cap aloft (reference chapter 4, section e) –around the 700-millibar level. The cap inversion is not so strong that it is unbreakable. When the cap breaks in a moist and very unstable environment, an explosive updraft results in rapid deep convection. In mid-latitudes, tops of supercells can exceed 60,000 feet and even 70,000 feet, the highest vertical growth of any type of thunderstorm. Thus, the local atmosphere in the low and middle troposphere must be extremely buoyant, allowing the air parcels to rise readily and rapidly. The environmental vertical temperature lapse rate is great enough so that rising parcels remain warmer than the environment they are rising through, well into the troposphere, and then as these parcels rise into the upper troposphere and very lowewr stratosphere and are colder than the environment, they are rising so fast that when they start to decelerate due to being negatively buoyant, they still rise to great height before their vertical velocity reaches zero.

We can conclude that a supercell is a highly-organized form of thunderstorm. Supercells represent only a small percentagea of all thunderstorms, but they are important because they are the most severe of thunderstorms.

Similar to the ordinary thunderstorm, the supercell has one main updraft, but its updraft can exceed 150 miles per hour. Since the supercell's updraft rotates, this characteristic allow us to distinguish it from other types of thunderstorms. ***The rotating updraft of a supercell thunderstorm is called a mesocyclone.***

The rain at the leading edge of a supercell's precipitation area is usually light, with the heavier rain occurring closer to the updraft, and with the torrential rain and/or large hail falling just north and east of the main updraft. The area near the main updraft, which is typically towards the rear of the storm, is where the severe weather occurs.

By way of review: a supercell is the most dangerous type of thunderstorm since this type produces most of the most violent, largest and longest-lasting tornadoes, most of the large hail events, and most of the strongest downbursts and microbursts.

The rotating updraft of a supercell may persist for a few hours. Even though most supercelss last for form about three to six hours, cases of up to eight hours occur.

What separates routine or ordinary thunderstorms from severe thunderstorms is the strength of their updrafts. In general, the stronger the updraft, the more likely is the storm to produce severe weather (a tornado, large hail and damaging straight-line winds or microbursts). In a supercell, the updraft rotates (a mesocyclone), which may lead to tornado-genesis.

The most severe of supercells form in an environment of great buouancy and large vertical wind shear (change of wind direction and or speed with height).

When an updraft stengthens in a vertically-shared environment, the air pressure is lowered on the updraft's flanks. The stronger the updraft and the stronger the horizontal air flow, the lower the pressure on the flanks. This develops increases the pressure change (pressure gradient) in the vertical, and when there is then lowering pressure with height, this acts to accelerate upward the air parcels from along the surface. This forcing helps to sustain the updraft, and also promotes storm propagation that deviates from the mean steering winds aloft.

Sometimes, part of a supercell will split off from the original storm and move to the right (in the Northern Hemisphere) and sometimes to the left, of the movement of the original supercell. The vertical wind directional shear and wind speed shear profiles help determine the initial movement of these right-splitting and left-splitting supercells, which are typically severe. The overwhelming majority of the time when a splitting cell occurs, it is a right-mover, going into the usually warmer, moister and more unstable air.

When the weather radar shows a supercell splitting away from the original supercell, this indicates that their associated updrafts may be rotating, and since rotating updrafts are mesocyclones which may produce tornadoes, the supercell thunderstorms and always potentially dangerous.

The wind direction usually veers (turns clockwise) with height and the wind speed increases significantly with height in right-mover supercell splitting episodes, and the wind direction usually backs (turns counterclockwise) with height and the wind speed increases significantly with height in the left-mover supercell splitting episodes. This is in the Northern Hemisphere; in the Southern Hemisphere, the directional shear directions are reversed.

miles ahead of the cold front, then the sun would come blaring out after the line passes and often rapidly redestabilizes the local environment so that after the pre-frontal storms pass, then the frontal storms move in perhaps two to six hours later.

The squall line is distinguishable from other storms or systems of storms by its continuity as a system. A squall line may be comprised of ordinary thunderstorms, multi-cell thunderstorms and supercells in any combination, as well as including smaller lines. A squall line or part of it may also evolve into a mesoscale convective system (MCS) (to be discussed later).

The most active part of a squall line is usually just behind the leading edge. Parts of the line grow while other sections decay. The parts with the heaviest precipitation are generally the most active line segments.

When a squall line has a wavy shape, it is called a *line-echo wave pattern (LEWP)*, which is dangerous because portions of the line may later intersect, generating rapid new and likely severe weather a the intersections of line segments, since the outflows converge, forcing rapid upward motion of air parcels.

In the Northern Hemisphere, the southern end of a squall line tends to have the greatest proportion of severe weather, since it is typically in the most unstable air while the rest of the line is in the process of stabilizing the local atmosphere. It is important to remember that a thunderstorm is not only a "thing", containing water vapor, but is also a "process". It is a dynamic machine whose job is to take a locally unstable atmosphere and mix it up so that it becomes stable.

All of the component cells of a squall line are competing for air inflow and moisture. When the low-level inflow becomes convergently concentrated at one part of the line, then that is where a particular cell or group of cells becomes stronger than the rest and possibly severe.

Although the "tail-end" cell of a squall line tends to have the greatest potential for rapid and possibly severe development, the genesis of severe storms may occur anywhere in the line when there are breaks in the line. These breaks somewhat reduce the competitiveness for inflow and moisture, and when the breaks occur during the daytime, especially on a hot and unstable day, then the solar heating of the ground can enhance convective development along these breaks.

Another potential condition for severe thunderstorm conditions form a squall line occurs when the line intersectrs another boundary such as another squall line, a front, an outflow boundary, a sea- or lake-breeze front, a dry-line, etc. The convergence and resulting lift from each boundary combine, resulting typically in rapid convection at the intersection with possible subsequent severe weather.

Storm development is also enhanced by low-level wind speed convergence. This is when an area has strong winds and an adjacent area has lighter winds. If the air flow is from the more windy area into the lighter winds area, then the air parcels "pile up" as they flow from a region of higher winds into the lower wind speed area. This convergence of air parcels forces them to rise. Thus, a region of low-level wind speed convergence can enhance the lift required for thunderstorm development, sustenance and intensification.

One of the factors that contribute to maintaining a squall line is the line moving through unstable environments. Another factor is that the gust fronts from individual cells components intersect to form new convection. Thus, a squall line is self-perpetuating as long as moist unstable air is encountered as it moves along. New convection forms on the line's forward flank as the boundaries intersect. Moreover, as was discussed earier, the tail-end of the line continues to have development of new cells on its right or right-rear flank, which results in a generally southward propagation of the end of the line.

The squall line propagates in some direction, but he individual cells usually move somewhat differently. Typical in the Northern Hemisphere is for a squall line to move generally southeastward while individual cells that comprise the line move northeastward; the resultant direction of storm is eastward, although an area some miles to the east will not be experiencing the same individual storm cell that was those same mile to the west earlier; also, the line, if an active one, is usually growing southward at its southern extremity.

The leading edge of the squall line, the gust front, scoops up the moist surface air and may form a roll cloud, which is a cloud that looks like a horizontal elongated roll, and may not be attached to the cloud base just above it. Also, some of the moist air in the outflow may be recycled into the updraft along the gust front, resulting in a lowered cloud base which is linear, and is attached to the thunderstorm cloud: it is called a shelf cloud.

Sometimes a roll cloud which is rotating may be tilted by inflow or an updraft to make it reach the ground. It looks somewhat like a tornado but is not a tornado; it is called a *gustnado*. There are usually gusty winds with these rotating, tilted roll clouds. In the Northern Hemisphere, virtually all tornadoes come out of the southwest part of a thunderstorm, not the leading edge, although rarely a tornado may form behind the shelf cloud.

Behind a squall line, the air is sinking, which causes clearing or clear skies.

Note: it is not uncommon for individual thunderstorms to develop ahead of a squall line, as well as for a squall line to exist within a larger area of convection.

Figure 14-6. A squall line as viewed from above, looking northeast to southwest. (source: FAA)

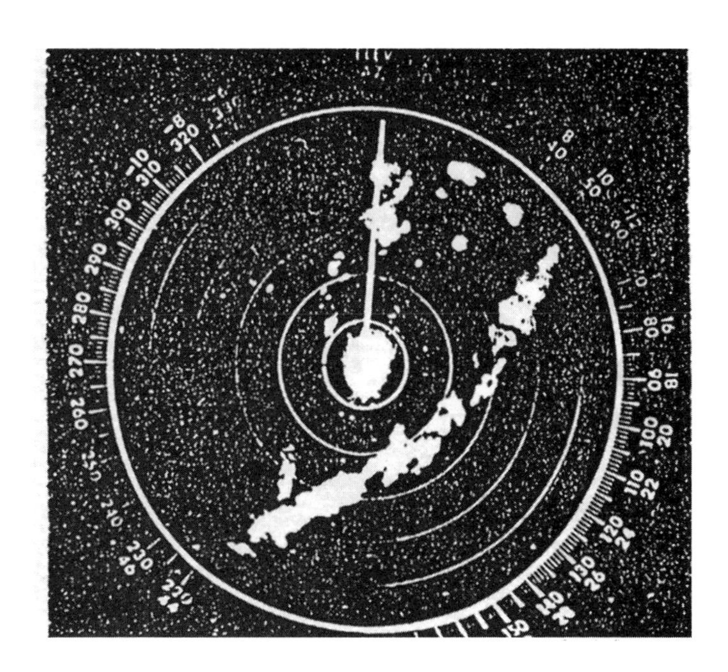

Figure 14-7. A radar scope showing a line of thunderstorms. In the middle of the scope is the radar site. Each concentric ring is 25 nautical miles (1 nautical mile equals approximately 1.15 statute miles) distance from the next ring. Notice the blob of echoes in the middle where the radar site is. This is called ground clutter, and is always there. It is the radar beam hitting and detecting buildings, hills, trees, etc. within several miles of the radar. (Source: NWS)

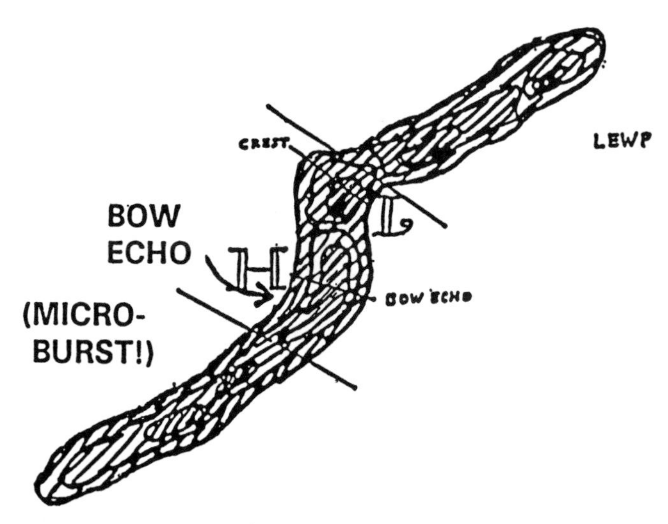

Figure 14-8. A line-echo wave-pattern (LEWP), which is a wavy squall-line of thunderstorms, and a BOW ECHO. **A LEWP is typically associated with severe thunderstorms. As parts of the line merge, new and severe storms rapidly form at the merging. Where the line bows in, we have rapidly descending air no echoes) and a possible microburst of winds downward from the clouds, often in excess of 100 miles per hour. On the back side of where the radar echo shows a "bowing out" may imply the rapid descent of air behind the rain echo, indicating a possible downburst of air, or a more concentrated downburst known as a microburst.**

Range-Height Indicator of a weather radar:
The radar antenna can keep rotating, sending out signals that look for precipitation to reflect off and try to return to the radar site. The radar antenna rotation can also be stopped so that the radar can aim in one direction at a precipitation area that may be worthy of detailed examination. For example, if a severe thunderstorm is rapidly growing, the radar can do a tilt-sequence, scanning up and down to see how high and how intense the echoes are. When high decibel levels are occurring in mid-levels of a developing thunderstorm, this heavy precipitation must plunge to the surface, meaning that such a signature may be indicative of a downburst of strong and damaging winds. Moreover, high decibel levels (very bright reflectivity on the radar screen) through a deep vertical range indicate that torrential rainfall may be occurring with that storm.

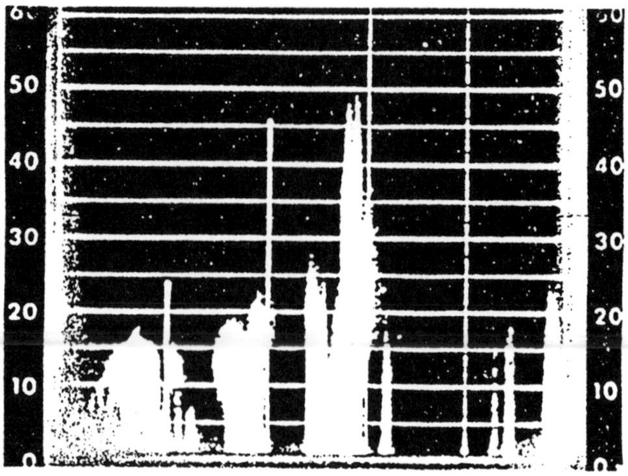

Figure 14-9. A display on the range-height indicator (RHI). The rotation of the radar antenna has been stopped and the antenna is scanning up and down, looking for the vertical extent and intensities of precipitation echoes. The vertical scale of the RHI display at right is labeled in tens of thousands of feet, and the bright areas are precipitation. One cell, for example, has precipitation tops to nearly 50,000 feet. The bright areas here are convective (shower, thunderstorm) cells. (source: NWS)

Figure 14-10. A roll cloud. This cloud is detached from the base of the cloud above it and sometimes occurs along the leading edge of a squall line. It is caused by the gust front, which is the squall line's leading edge, scooping up some of the moist surface air. (source: NOAA)

Figure 14-11. A shelf cloud. This cloud is attached to the thunderstorm cloud, and is formed when some of the moist air in the outflow is recycled into the updraft along the gust front, resulting in a lowered cloud base which is linear. (source: NOAA)

f. derecho

A derecho is a convective cluster of downbursts. This storm may develop from another type of thunderstorm.

Derechos are thunderstorms with violent straight-line winds (downbursts). In the United States, they tend to occur with a northwest flow, with the geographically-favored region being east of the Rocky Mountains.

A huge amount of energy must be expended to generate damaging winds of about 80 miles per hour (130 kilometers per hour) or greater, which can persist for hours as this cluster of storms moves rapidly through. To generate such energy, huge amounts of moisture must be usable in the form of the release of latent heat of condensation as the air parcels rise and the water vapor condenses into cloud matter and precipitation. We believe this is the main driving force for such storms since they typically form only in regions with concentrated very high dewpoints,.

For example, consider northern Wisconsin, which is a prime area for this infrequent event. In the heat of summer, the dewpoints in this region infrequently rise to almost and sometimes around 80 degrees Fahrenheit. This is because the air is densely forested, and in the summertime on a sunny hot day, a mature oak, maple or pine tree can release from about 50 to 100 gallons of water vapor into the boundary layer (about the lowest 1500 to 2000 feet of the atmosphere during the daytime). This region is densely populated with mostly large pine trees. Thus, an enormous amount of moisture is made resident in the local area, which, when the other conditions (or "ingredients") for convection occur, result in thunderstorm development which may become a derecho thunderstorm system.

Thus, when dewpoints in a local area are about 79 or 80 degrees F. (about 26 degrees Celsius) or even higher, and the surrounding area has lower dewpoints, then this dewpoint pooling of copiously moist air is a region to watch for possible derecho development when the conditions exist for thunderstorms to form.

The term, "derecho" comes from the Spanish, meaning "straight ahead" or "direct", since this storm cluster of violent wind cells plows straight ahead rapidly, creating a swath of damage and destruction. Some derechos in southern Canda and in the United States (all east of the Rockies) have winds of 100 miles per hour (160 km/hr) or more that continue for ten or more minutes as the system moves though.

A derecho can be some 100 miles or even somewhat more wide, and can charge through a 100-mile or more swath; indeed, some can travel for hundreds of miles before diminishing. They persist longer when they are traveling through regions with high plenty of warm, moist air.

The flow aloft at some 500 millibars (about 18,000 feet) is typically northwesterly, and sometimes westerly flow in derecho episodes, so that these systems tend to be steered to the southeast or east.

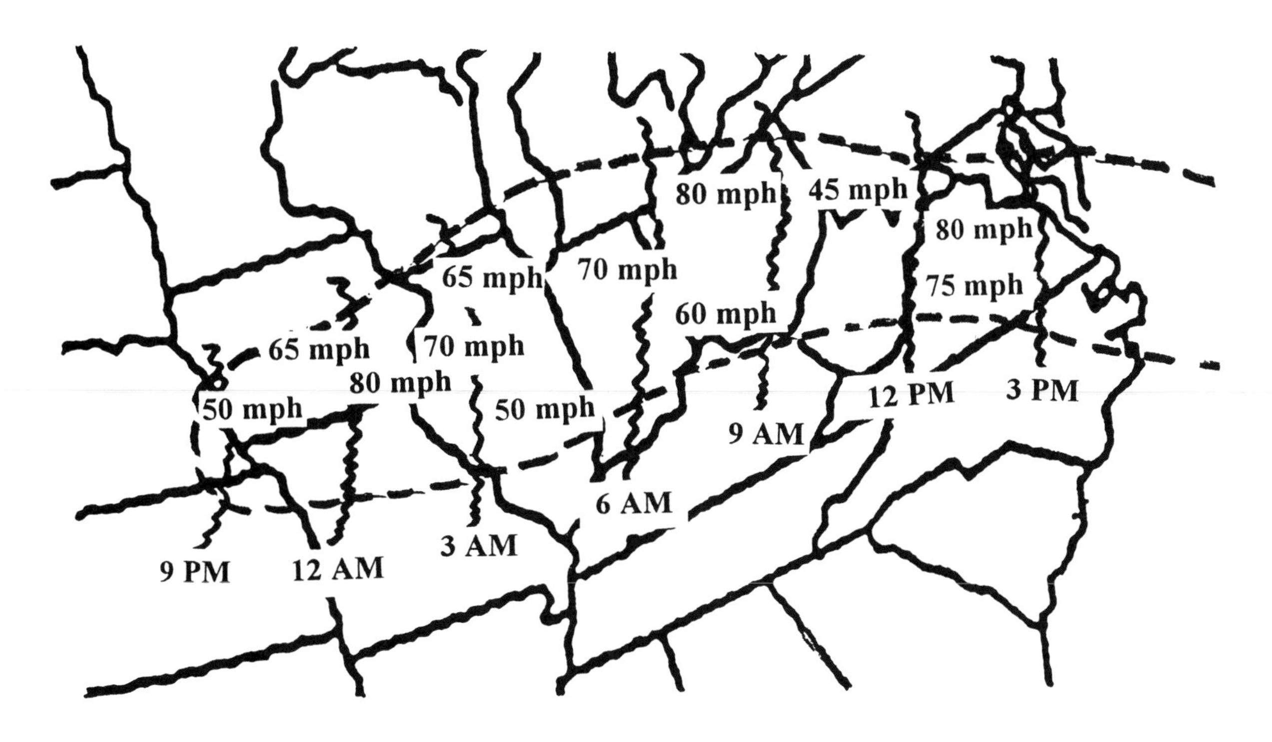

Figure 14-15. The path and width of one of the longest derecho episodes recorded. This event occurred in early July. The times are in central standard time for the locations of the central part of the derecho multi-cellular system.

g. dry-line thunderstorm

A phenomenon that occurs during the warmer part of the year in parts of the world where a large, warm body of water is adjacent to a large land mass that becomes semi-arid well-inland, is the dry-line. The dry-line separates warm, moist air from dry air. Thus, this boundary acts as a type of “front”. The dry-line moves back and forth throughout the day. During the hot daytime, the air on the dry side is hotter than on the moist side. This is because as the air contains more water vapor, less of the shorter-wavelength solar radiation makes it to the grund to heat it up so that the air above it is heated by the radiation that the earth then emits as longer wavelength infrared radiation. One way heat is carried is through infrared radiation. Inotherwords, when the air is more moist, the day’s high temperature is held down somewhat. The water vapor absorbs and reflects some of the solar radiation, and although the water vapor itself warms up, the overall warming of the local environment is less than it would be in drier air.

For example, with everything else being equal, a summer day that starts out with a dewpoint of 55 degrees with sunny to mostly sunny skies expected and no cold air moving in, may see the high temperature reach 100 degrees Fahrenheit, but a day starting out with a dewpoint of 70 degrees may see a 93-degree high.

In the United States, the warm-season dry-line typically extends from Kansas through Oklahoma and Texas. It can extend as far north as Nebraska. The dry-line "front" moves westward as far as eastern New Mexico and eastern Colorado, and moves eastward as far as eastern Kansas and western Missouri and to eastern Oklahoma and towards eastern Texas.

In the summer, it is common for the dewpoints in east Texas on the moist side of the dry-line to be in the 70s while the dewpoints in west Texas on the drierside of the dry-line to be in the 40s and 30s. However, the actual air temperatures in west Texas during the daytime are typically in the 95 to 110 degree range, whereas in east Texas they are in the 90s, and at night the temperatures in the west are cooler thanin the east when the skies are clear or mostly clear.

The importance of the dry-line is that convection often develops along it. When the dry-line boundary intersects another boundary, such as outflow boundaries from non-dry-line thunderstorms or intersects with an advancing cold front, then thunderstorms often form which can be severe.

Often the appearance of a dry-line thunderstorm is bell-shaped, and striations within the cloud can be seen as the updraft winds to the top.

On many warm-season nights in Texas and Oklahoma, the air cools off rapidly at the surface, resulting in a pocket of warm air just above this. This is a capping temperature inversion. The inversion is usually broken by the mixing of air due to intense solar heating and the breezes that occur. This mixing of the drier air aloft with the surface air causes the dry-line to show an apparent movement eastward. If the dry-line encounters another temperature inversion, it stops advancing, but the inversion must be sufficiently strong to stop the line's movement.

Eventually, as the daytime heating slows, so does the eastward progression of the dry-line. By nightfall, the low-level temperatures drop faster in the drier air to the west of the line than in the moist air to the east, which generates a pressure gradient (pressure difference) across the dry-line. Lower pressures develop on the west side; therefore, the lower-level winds have an east-to-west component from the higher to lower pressures across the boundary , and the dry-line progresses westward at night. Thus, *typically, the dry-line advances eastward into the moist air during the daytime, and moves westward into the drier air at night.*

Eventually, at night, the dry air side jof the line cools to the point at which it is denser than the moist air to the east, and the boundary ceases progressing westward. The dry-line therefore becomes quasi-stationary until daytime heating commences anew.

This scenario is modified under the influence of other weather systems, such as a strong cold front, tropical cyclone, etc., but in a weak summertime pattern, the dry-line's behavior is typified by the above discussion.

Analysis of weather data by vertical soundings such as from radiosondes shows that the farther east one goes into the moist air, the higher the level of the capping inversion. Thus, this slope causes the dry-line to resemble a front. Thus, a dry-line can be misanalyzed as a front on a weather map of that region.

Figure 14-16. A typical dry-line thunderstorm is frequently bell-shaped to some extent, and often has visible striations in its cloudmass; thus, you can "see" the curving updraft of the storm. Moreover, many dry-line thunderstorms rotate.

The updraft of a dry-line thunderstorm is usually one large cyclonically-rotating rising draft, although on rate occasions it rotates anticyclonically. Rainfall amounts are not particularly heavy, although large hail can occur. The hail usually falls upwind of the main cloud tower, out of the high-level outflow.

Even when dry-line thunderstorms become severe, their anvils are typically not extensive.

Most of the precipitation falls ahead of the updraft, probably because the dry-line storm has no organized downdraft, particularly at the rear flank.

When the line moves into moister air, it is possible for dry-line storms to evolve into other types of thunderstorms, such as multi-cells and supercells.

An often explosive scenario for severe thunderstorms in this region of the country is when the dry-line intersects with another boundary, such as a frontal boundary, an outflow boundary from other storms, a spiral band from a tropical cyclone, etc. But even such mergers, dry-line thunderstorms on their own can become severe, producing large hail and big tornadoes. Indeed, a dry-line storm, having a rotating updraft, has a characteristic of a supercell. However, in the case of the supercell thunderstorm, the tornado is typically found around the updraft/downdraft interface, whereas there is virtually no concentrated, well-organized downdraft ina dry-line thunderstorm.

If you want of consider the dry-line thunderstorm as a subset of the supercell classification, then another major difference between the two is that the supercell produces heavier precipitation than the dry-line storm.

Also, it is possible to have other types of thunderstorms form along the dry-line boundary.

h. mesoscale convective system (MCS)

A mesoscale convective system or MCS is a large, organized area of convection that produces very heavy to excessive rainfall, with the system persisting for up to twelve hours or more. MCSes form only in regions of concentrated warm and moist air, where the other ingredients for convection are also present. These regions of concentrated warm, moist air are called theta-e ridges, which are explained in detail in chapter 23.

Most of the flash floods are caused by MCSes, which is why forecasting their development and movement is crucial.

An MCS can be as big as the state of Iowa. Unlike most air mass thunderstorms, which move in the direction of the air flow around 12,000 feet off the ground, MCSes tend to stay in the same type of vertical thermal environment they formed in, i.e., to move along with the thickness pattern. Thickness is how thick is the air between the surface and 500 millibars. The MCS tends to move in the same thickness ribbon except for the case in which the thickness value lines diverge (called difluent thickness), in which case the MCS propagates backwards to where the thickness lines are closer together. For our readers who are not meteorologists and for whom this terminology is unclear, what this means is that the MCS essentially prefers to propagate towards the region that has the same essential temperature conditions in the vertical in which it formed.

Figure 14-17. A mesoscale convective system (MCS), which is a relatively large blob of organized thunderstorms, is indicated by the M. The plume of moisture from the tropical Pacific Ocean is called a ***tropical connection***, and makes the very heavy precipitation from an MCS even heavier, since it continuously "seeds" the MCS with additional moisture to produce precipitation. (source: NOAA)

An MCS can easily dump three to five inches of rain over an area as it passes through, although it takes several hours for an entire system to pass overhead, and rains of over five inches are common enough.

A decaying hurricane over land becomes an MCS, still producing very heavy to excessive rainfall. Occasionally, an MCS may work its way to over a warm body of water such as the Gulf of Mexico, subsequently evolving into a hurricane.

In the United States, MCSes occur most often between the Rocky Mountains and Appalachians, and are responsible for a considerable percentage of the rainfall taht falls in this region during the growing season.

Occasionally, an MCS, which is typically oval-shaped, forms where the upper-level winds are at that time very light, and forma s a circular or nearly-circular system. These MCSes tend to produce even more rainfall than oval-shaped MCSes since the light winds aloft mean that they move more slowly and thus spend more time dumping very heavy rainfall over an area. Sometimes these circular MCSes are referred to as mesoscale convective complexes (MCCs).

i. hurricane

A hurricane is a large storm that forms in the tropics and sometimes just outside the tropics, having sustained winds of at least 74 miles per hour (about 120 kilometers per hour). The general name for these storms is tropical cyclone. The name, "hurricane", is used in the Atlantic and eastern and central Pacific (coming from the Carib Indian word "huracan"), whereas in the western Pacific basin, the term used is typhoon is used. Around Australia the term used is "tropical cyclone", and in the Indian Ocean area the terms "cyclone' and "tropical cyclone" are used.

The hurricane is a normal aspect of the earth's climate. If we were able to destroy hurricanes before they made landfall, Nature would have to generate another means to transport the net build-up of tropical energy towards cooler latitudes. Thus, the function of the hurricane is to contribute to the moderating and regulating of the earth's climate. Because these storms are so significant, hurricanes are given names for historical reference.

Hurricanes can grow to be as large as several hundred miles across, but the winds get stronger as we approach the center of the storm. In the center itself, the wind is nearly calm.

The air pressure in and around the center is quite low; only the relatively tiny tornado can have lower central pressures. Pressures have fallen below 26.00" on the barometer, close to 850 millibars (hectoPascals) in the most powerful tropical cyclones. For example, in October 1979, Typhoon Tip in the western Pacific has a pressure of 25.69".

Around North America, most hurricanes occur between June 1st and November 30th, with the peak of the hurricane season being between August 15th and October 15th. September is the month with the most hurricanes. In the eastern Pacific, the season typically starts about a month earlier. In the western Pacific, typhoons can develop during any month. In the Southern Hemisphere, tropical cyclones are most likely from January through March.

Hurricane development:
Areas of thunderstorms in the tropics (and sometimes just outside the tropics) often organize into a complex of convection. This is a region of converging air, which rises through most of the tropopause. These blobs can form initially over land, such as over equatorial Africa, and move into the Atlantic Ocean, or they can form over the ocean. About one-fourth of all tropical storms form in the Intertropical Convergence Zone (ITCZ), which is an area of converging winds in the tropics.

During the hurricane season, there are about 100 seedlings that have the potential to become tropical storms. An average of 10 to 12 of them make it every year. When the highest sustained winds in the system reach 39 mph, the storm is then called a tropical storm. If the storm's highest sustained winds reach the threshold value of 74 mph, it is then termed a hurricane. Out of the 10 to 12 tropical storms, about 6 to 8 of them grow into hurricanes.

Most seedlings are called easterly waves since they initially typically move westward through the tropics. Such systems are initially called tropical depressions. Tropical depressions also form out of squall lines that move into the ocean from Africa, and from remnants of other weather systems that have some vertically rising air to them and have been lingering for several days (such as the remnants of an old cold front that makes it into or near the tropics).

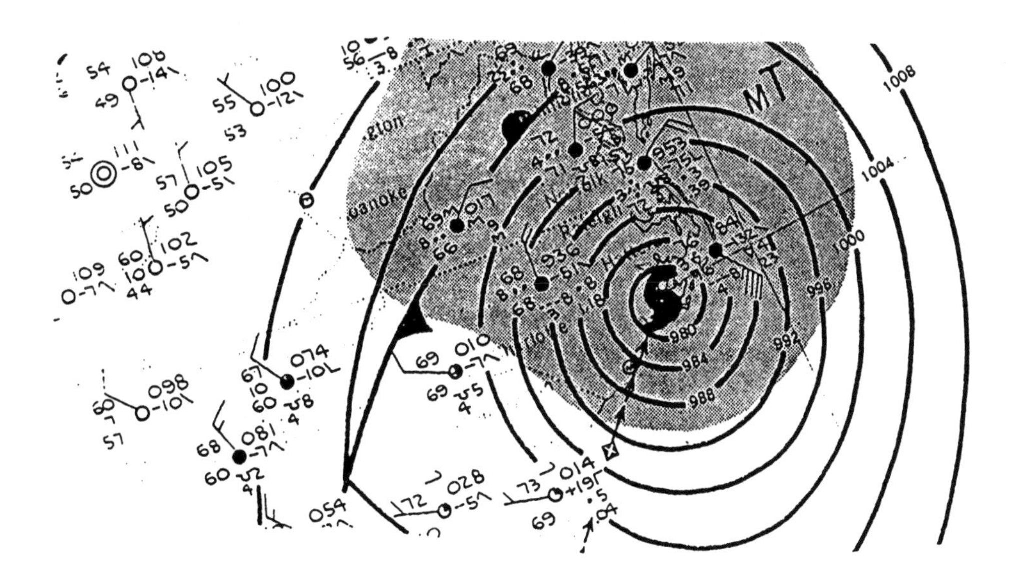

Figure 14-18. How a hurricane is depicted on a surface weather map. Notice how tightly wound-up the isobars, or lines of equal pressure, are, especially as we get closer to the center of the storm, which is called the eye. The shaded area is where rain is falling. (source: NWS)

Now, let us take a detailed look at the structure of a hurricane.

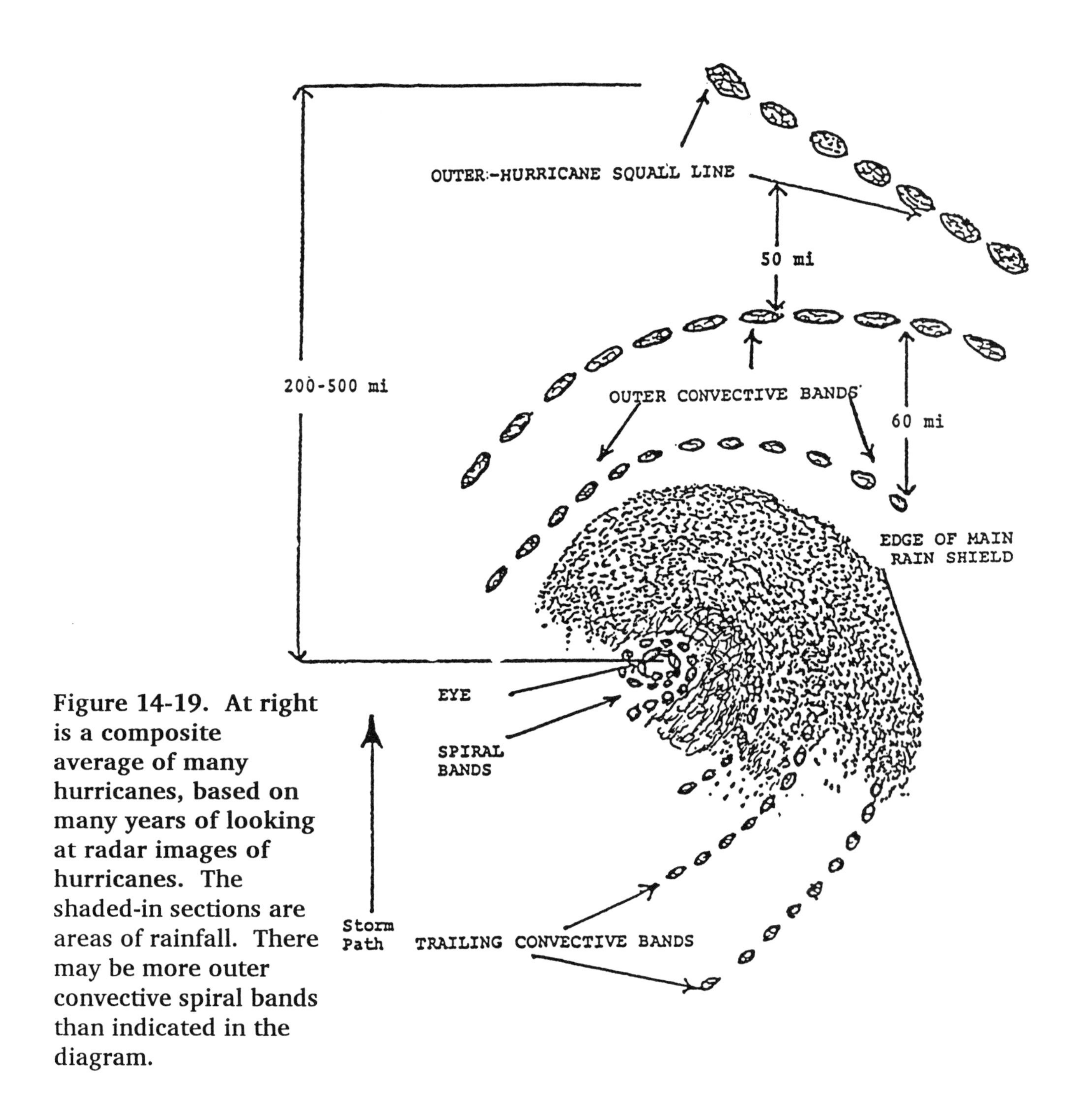

Figure 14-19. At right is a composite average of many hurricanes, based on many years of looking at radar images of hurricanes. The shaded-in sections are areas of rainfall. There may be more outer convective spiral bands than indicated in the diagram.

Figure 14-20. Here is a weather satellite picture of a major hurricane (Hurricane Gilbert) west-northwest of Jamaica. Notice the eye, cyclonic circulation, feeder bands and anticyclonic outflow aloft. (source: NHC)

Hurricanes require sea-surface temperatures of at least 79°F (26°C) to flourish. If another weather system has strong winds aloft in such a direction as to shear off the tops of the tropical cyclone, then the tropical system will fall apart.

<u>The maintenance of the hurricane</u>:
Evaporation of some very warm surface water provides the warm, moist air which contains much of the energy potential for hurricane development. As this air rises into the storm, it cools to its dewpoint, forming clouds, and as it keeps rising, cloud droplets coalesce and accrete to form the raindrops. In this condensation process, the "heat of condensation" is released. This is a form of energy. The hurricane uses this energy to maintain itself and to grow.

When the temperature and dewpoint are high, as over the tropical waters, the energy available is enormous. One study concluded that the amount of energy expended by a typical hurricane would, if it could be harnessed, supply the entire energy needs for the United States for half a year!

<u>The air circulation in a hurricane</u>:
Air spirals into the hurricane, rising as it does so, encircling the relatively calm eye. In the eye, the air is sinking, warming adiabatically as it descends. Although the perfect model of the circulation and energy evolutions of a tropical cyclone has not been formulated, we do have sufficient observations and data to suggest some circulation models of the storm. Much of the air is also recycled within the storm. There is vertical exchange of energy, moisture and momentum. There are smaller-scale, called mesoscale, features within the hurricane. For example, not all hurricanes should be expected to have vertical eye-walls. Indeed, one study of five hurricanes found that their eye-walls leaned outward, which would suggest a sloping updraft.

Some of the details that a comprehensive hurricane circulation model should account for are:
- the maintenance of the hurricane's convection;
- the lifting of evaporated water out of the planetary boundary layer;
- the greater tapping of the ocean energy source through downdraft drying and cooling of the boundary layer; and
- the balancing of the hurricane's circulation against radiational cooling.

<u>Note about the anticyclonic outflow aloft</u>:
The air does not just keep spiralling in counterclockwise in the Northern Hemisphere and clockwise in the Southern Hemisphere, and disappearing. Due to conservation of mass, the air has to go somewhere to maintain a circulation. Aloft in the hurricane, towards and at and over its top, the air comes out in an anticyclonic (clockwise in the Northern Hemisphere and counterclockwise in the Southern Hemisphere) fashion. Thus, a hurricane moving across the Atlantic towards the United States, and a typhoon heading across the western Pacific towards Japan, and a tropical cyclone heading across the Indian Ocean towards India, all have, since they are Northern Hemisphere tropical cyclones, a counterclockwise cyclonic circulation with an anticyclonic outflow aloft. This outflow is hard to distinguish on individual weather satellite pictures, but on time-lapse video-loops of these pictures you can detect a surge of high cirroform clouds spiralling clockwise aloft out of the storm.

Violent winds and flash flooding rains are not the only onslaught from a hurricane. The storm surge is the greatest killer in a hurricane. A storm surge killed at least 6,000 people...perhaps as many as 12,000...in Galveston, Texas in September 1900, and killed upwards of some 500,000 people in Bangladesh in November 1970.

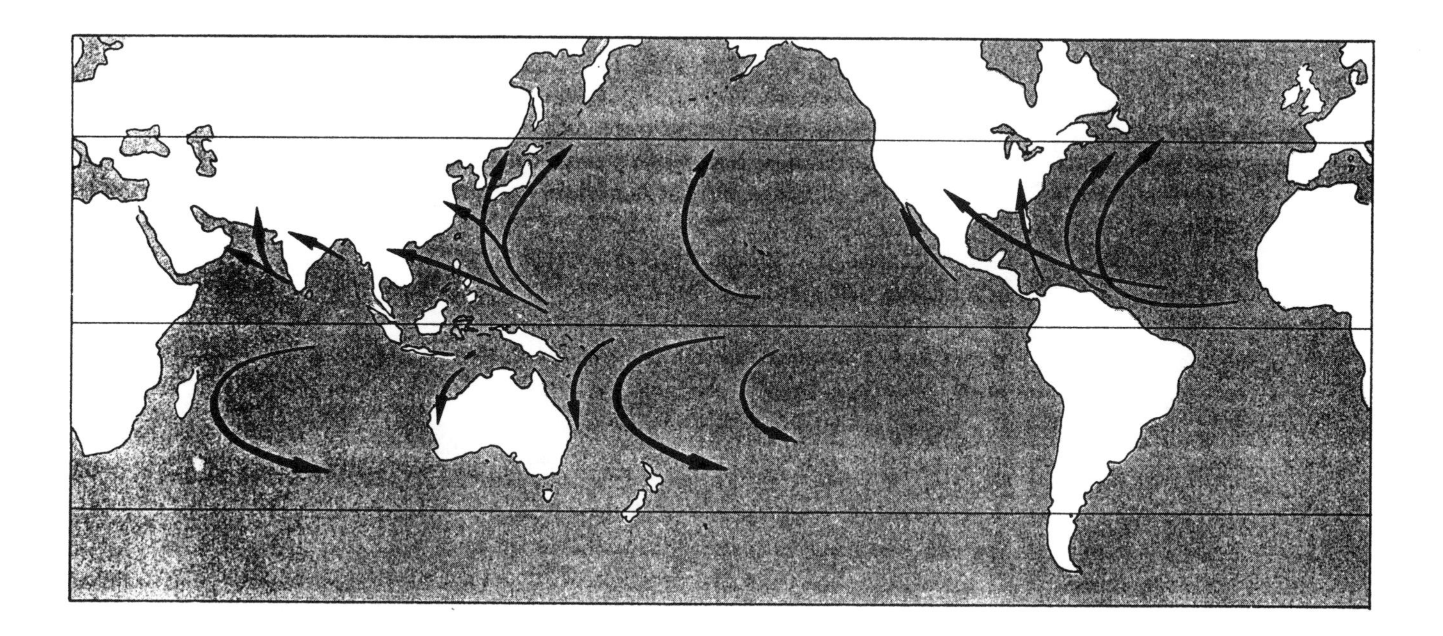

Figure 14-21. The MAIN (not all) areas of tropical cyclone formation and favored storm tracks. In the North Atlantic Basin, most tropical cyclones occur in August, September and October. (source: USAF)

For a detailed explanation of hurricanes and typhoons, please refer to the book, *"HURRICANES!"*, written by this author in the same format and style as is this book. *"HURRICANES!"* is available from Chaston Scientific, Inc. at the address on the back cover.

(This author is indebted to work done by various research facilities of NOAA, the National Oceanic and Atmospheric Administration, for much of the latest knowledge on the various thunderstorm types discussed in this chapter. Information derived from this research has been incorporated into this chapter.)

Chapter 15. HAIL

Figure 15-1. A violent hailstorm with large hailstones covering the surface during a severe thunderstorm.

Hail is not sleet. Whereas sleet occurs typically in stratiform clouds such as stratocumulus, hail forms in the cumulonimbus clouds of thunderstorms, whose cloud tops can grow into the lower stratosphere. Meteorological research has still not proven conclusively how hail forms. We do know that when large hailstones are cut in half, we observe layers of ice, similar in appearance to layers of an onion. This implies that hail formation involves an ice accretion process.

Early theories speculated that the hailstone must be undergoing successive rides up and down through the thunderstorm, going above and below the freezing level, propelled by updrafts acting perhaps in a gusty fashion. Each time a hailstone would fall below the freezing level, it would pick up another coating of water, and if carried by gusty updrafts to above the freezing level, that water would subsequently freeze, adding another layer of ice to the hailstone. The hail would then perhaps undergo the process again and again. Ultimately, the hailstone would fall to earth only when it became too heavy to be supported by updrafts or the updrafts weaken.

This theory also assumed that the hailstones formed around the freezing level in thunderstorms, which on a summer day in mid-latitudes could be 12,000 feet or higher above the ground, and in spring, often from 6,000 to 9,000 feet elevation. What vertical depth the hailstones travel in their creation process is unknown. Some researchers proposed that the stones may not travel up and down too far, but would keep accreting layers of ice around the freezing level in the thunderstorm cloud.

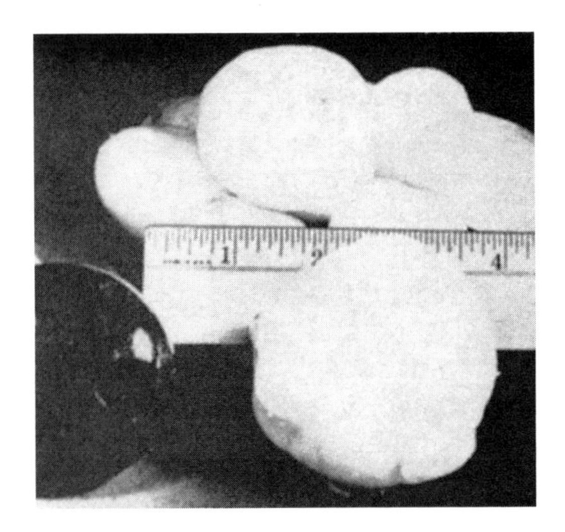

Figure 15-2. In the United States, the worst hailstorms occur in the central plains. Compare the size of these hailstones with the doorknob in the left of the picture.

A later theory of hailstone formation proposed that the hail forms as a frozen raindrop or partially-melted or totally-melted snowflake that refreezes, being held up in the cloud by an updraft, during which time the hailstone has supercooled drops of water adhering to it and freezing, forming new ice layers. (Supercooled water is water existing in the liquid phase even though the temperature is somewhat below freezing. We know that this can occur in clouds.)

The popular theory now is that some small object needs to serve as the nucleus for hail. This object could be a large, frozen raindrop or, because small objects such as insects have been found inside the middle of hailstones, even insects could serve as hailstone nuclei. The nucleus then accumulates supercooled liquid droplets (cloud droplets) through accretion. Updrafts carry the hailstone through some portions of the cloud, and through varying liquid concentrations. When the layers of ice that form around the nucleus cause the hailstone to be too large to be held aloft by updrafts, or if the hailstone is flung out of an updraft, then, thanks to gravity, and maybe also to a downdraft, it heads for earth.

The actual cause of hail probably embodies at least part of all these theories.

Many small hailstones melt into rain before reaching the surface. Also, hailstones can have layers of clear and opaque ice.

The largest hailstones have been over five inches (13 centimeters) in diameter. In 1995, Chinese weather bureau meteorologists reported basketball-size hail in one of China's provinces. Such enormous hail killed people and demolished homes and crops. Sometimes hail falls for over ten minutes and accumulates, even forming hail drifts when pushed by flash-flood flowing water. These hail drifts, which can be several feet deep, can persist for a few days, even in summer, before totally melting. Grapefruit-size hailstones require updrafts on the order of 60 to 70 meters per second (about 125 to 150 miles per hour). Imagine being in an aircraft flying through such an updraft!

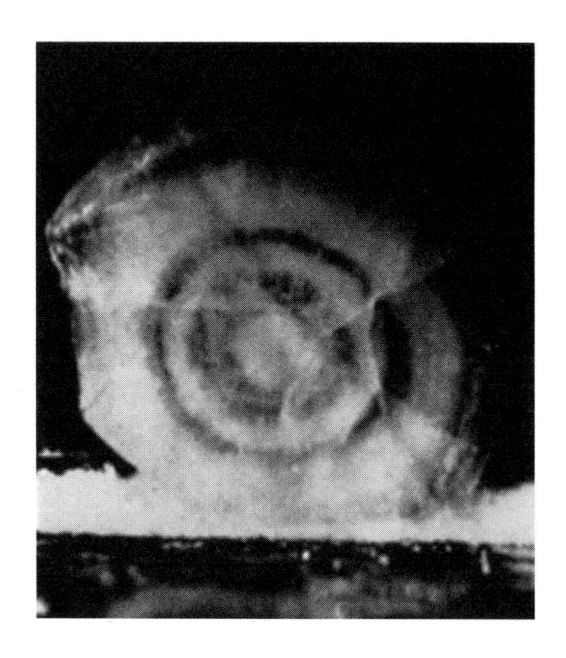

How to preserve large hailstones:

Figure 15-3. A large hailstone, cut in half to show the layers of ice.

If you experience large hail such as the hailstone in this picture, and you want to temporarily save some of the largest hailstones to show them to others and to take pictures and video of them, do not simply place them in a freezer, especially a frost-free freezer, since the ice will eventually pass into the vapor phase, causing your 3-inch hailstones to become pea-size and then disappear. Instead, place the hailstones in a plastic-like bag or container and seal the bag or container tightly, and then put them in the freezer. This prevents the sublimation from ice to vapor, and will therefore preserve your hailstones.

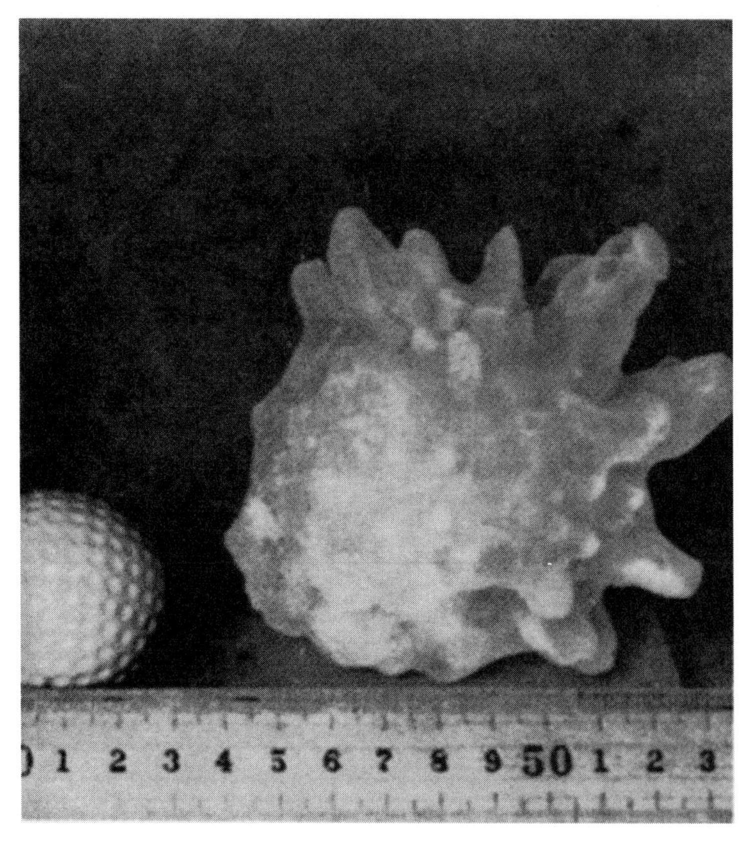

Figure 15-4. Not all hailstones are ball- or egg-shaped. Here is a photo of a large spiked hailstone that fell at Washington, D.C., a place not noted for large hail.

Figure 15-5. Large hailstones fall at speeds greater than 100 miles per hour (160 kilometers per hour) by the time they reach the surface! (source: NCAR)

Fig. 15-6. When hail persists for several minutes or more, it can eventually cover the ground. (source: NOAA)

Here are some eye-witness observations of some noteworthy hail events.

At about 9 a.m. local time on March 12th, 1975, a single piece of hail fell in the north part of Memphis, Tennessee in the midst of a thunderstorm with small hail. This one large piece crashed through the roof of a house, landing on the floor inside. The homeowner stated that the hailstone was 12 inches in diameter. If this hailstone could have been preserved, it would be the world's record. However, it is possible that this ice chunk could be a coalescence of several partially melted and then fused hailstones. Similar stories have been reported. For example, in Heidgraben, Germany, one hailstone was reported to be six inches by ten inches (which probably was fused hailstones) and weighed about four pounds when it crashed through the roof of a house. At Cazoria, Spain, a hailstone weighing 4½ pounds was reported. From India have come claims of hail "as big as a millstone", weighing 7½ pounds each. Parts of southern Russia sometimes receive hail so severe that it accumulates in the late spring and summer to depths of several inches and sometimes over a foot.

People sometimes are killed by large hail. In another personal yet relevant story, here is a hail experience that happened to this author. I was standing on the driveway in front of my house in rural northwest Missouri, watching very dark clouds of a thunderstorm forming directly overhead. The developing storm, however, was moving northeastward, and away, and as it did, some of the clouds developed an eerie greenish color, which occurs often in storms with large hail. Suddenly I heard a loud thump right beside me. Something had plunged from the sky just past my right ear and crashed into the ground. I turned to see what it was, and discovered a solitary large hailstone, resting on the ground all by itself. I thought to myself, "Just a few inches to my left and this big hailstone would have hit me right on my head, possibly giving me a concussion, even killing me. Then the hail would melt and evaporate, and nobody would have been able to find the cause of death." The storm did hit the next community about seven miles away with baseball- to softball-size hail.

But people do sometimes lose their lives because of hail. A farmer near Lubbock, Texas was caught in his field, far from any shelter, and was pelted so fiercely by large hail that he was killed by it.

Six children were killed and ten were injured by "hen's-egg-size hail" at Klausenberg, Romania. At Rostov, Russia, large hail pelted 23 people to death as they tried to save their cattle. In the Union of South Africa, nearly 20 people died when large hail buried them to a depth of three feet.

A widespread hail event occurred in the region around Hyderabad, India, ravaging 17 villages across 30 square miles. The hail was accompanied by very high winds which thrust the hailstones into houses and stripped leaves from the trees.

Hail is a popular subject even in ancient literature. It is also mentioned several times in the Old Testament. For example, the Book of Joshua tells of great hailstones killing the Amorites.

In this country, an infant was killed when struck by large hail around Ft. Collins, Colorado.

On July 31st, 1985, moisture from air that came from over the Gulf of Mexico was swept unusually far to the northwest, reaching into southeast Wyoming. A large thunderstorm developed, moving westward over Cheyenne. A tornado came out of the western part of the storm, and a flash flood then ensued. One elderly woman took shelter in her basement when the tornado warning was sounded, only to die of drowning from the raging flash flooding waters that rapidly filled her basement. Over 6 inches of rain fell in a little over 2 hours, which for that semi-arid region was excessive.

Part of Cheyenne was receiving a heavy accumulation of hail from the storm, and when the flash flooding waters reached the area wherein the hail had accumulated, it pushed the hail into drifts from 10 to 15 feet high, and also pushed the hail into some homes, filling the basements with hailstones. Even though this was a summer event, high drifts of hail were still in the area two days later.

Although most hailstones, especially small hail, are more-or-less oval, hail can fall in different shapes. Hailstones once fell in Italy in the shape of quartz crystals, and photographs of hail exist which show spokes of ice coming out of the hailstones.

Sometimes, hail can fall in great magnitude and last sufficiently long to accumulate to depths of over a foot. Hail drifts of six feet occurred in Washington County, Iowa.

Large hail is obviously dangerous to people and animals. It flattens crops. It damages buildings. One of the criteria for the issuance of a severe thunderstorm warning is for hail to be three-quarters of an inch in diameter or larger.

If you are driving your car and run into large hail, try to find an overpass under which to pull over to the side to avoid dents and a broken windshield. Grapefruit or softball size is the typical size of the largest hailstones, with rare exceptions of larger stones.

Figure 15-7. Compare the size of this enormous hailstone with the egg beside it. When hail of this relatively great size occurs, it falls in a hailstorm of mostly smaller hailstones, with some of the hailstones being of the greatest size. Hail the size shown here will pulverize crops, badly dent automobiles, can kill humans and animals and can even come through the roofs of some buildings. (source: USAF)

Hail is a hazard for aircraft. According to the U. S. Air Force, hail has been encountered by aircraft flying as high as 45,000 feet (13.7 kilometers). Moreover, sometimes the hail has been encountered, even at that altitude, in clear air to the side of the thunderstorm as some of the stones are flung out. The Air Force also reports that some of their pilots have encountered hail up to ten miles (16 kilometers) downwind from the storm core.

Hail larger than one-half to three-quarter inch in diameter can also damage aircraft. Aircraft on the ground that are not in hangars are particularly vulnerable.

As for crops, in the United States alone, some crop losses are to be expected from hailstorms. The region with the most hail days is from eastern Wyoming and eastern Colorado eastward through the plains states.

The United States is not alone in receiving hail, including damaging large hail. And hailstorms are not entirely restricted to over land; hail does occur in some thunderstorms over the oceans.

Hail occurs occasionally in the tropics, but usually at higher altitudes. The high freezing levels in the tropics may account for this, and the updrafts must rise to high levels for hail to even occur. Moreover, the hailstones fall through a deep depth of above freezing air and would most often melt before reaching the surface.

When hail occurs in polar regions, it is typically small hail.

Besides the United States, large hail can also occur in Canada, across Europe and Asia and in Australia. It can also occur elsewhere, but those regions with the most thunderstorm activity are obviously the most hail-prone areas of the globe.

Hurricane-hunter reconnaissance aircraft occasionally encounter small hail aloft while flying through hurricanes (according to a personal conversation with Robert Sheets, former Director of the National Hurricane Center who flew some 200 missions into hurricanes).. Hail being observed on the ground during a hurricane is rare. Small hail occurred briefly during the passage of the eye-wall of Hurricane Hugo at San Juan, Puerto Rico's airport on September 18th, 1989. It may occur somewhat more often than reported, since the fury of the blinding wind-swept rains may make hail difficult to observe. As a hurricane moves inland and its surface wind diminish, the winds a few thousand feet off the surface remain powerful for many hours yet, diminishing more slowly than the surface winds. This creates a wind-speed shear that makes the environment more conducive for severe thunderstorms, including hail-producers. This is especially true if the hurricane is moving inland into air that is already warm, moist, unstable and with no cap inversion aloft.

Sometimes when large hailstones are cut open, insects are found inside. These bugs were entrained into the thunderstorm and while its hail was being produced.

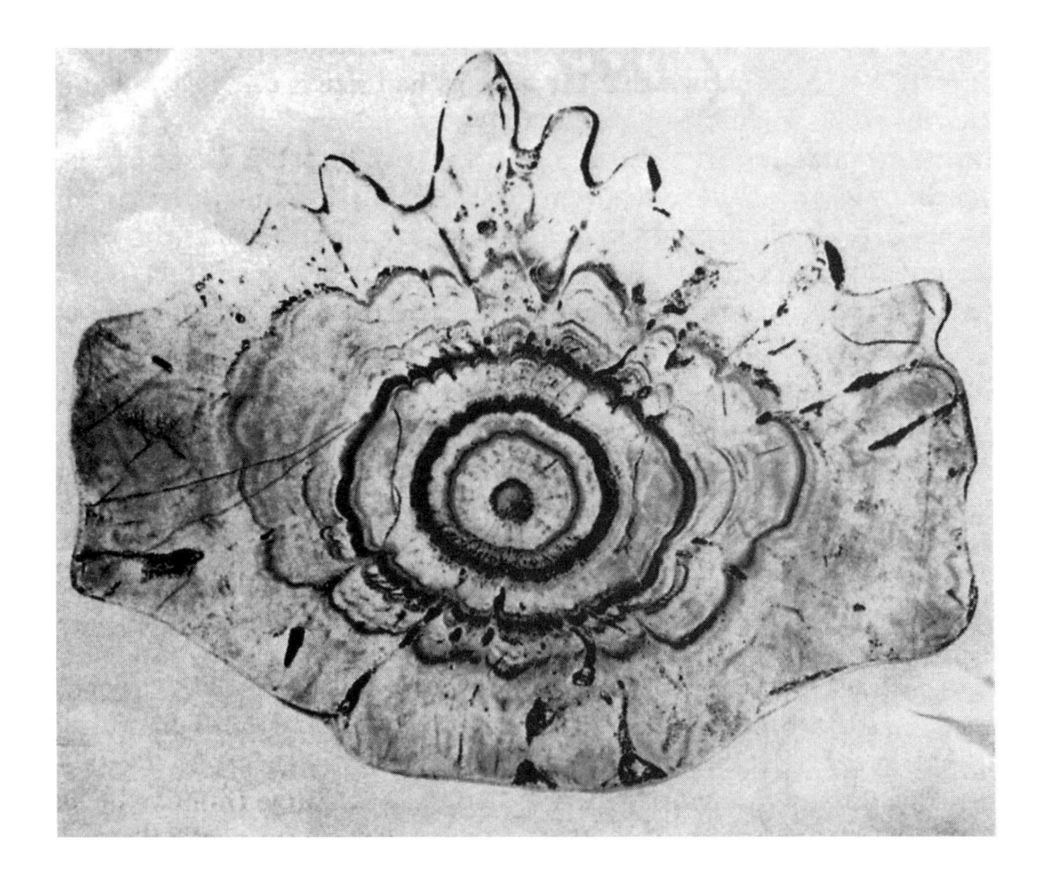

Figure 15-8. A photograph of one of the largest hailstones that ever fell, with the stone cut in two to show the layers or rings of ice. This hailstone crashed to the ground near Coffeyville, Kansas with a diameter of about 7½ inches (about 19 centimeters) across! Imagine having one of these conk you on the head or crash through the windshield of your car. (source: NCAR)

Typically, when large hail falls, the hailstones are of varying sizes. Only some of the hailstones will be of the largest size when the extreme (baseball-size or larger) hail occurs. In the United States, the National Weather Service considers the size of the largest hailstone as one criterion for the issuance of a severe thunderstorm warning. The largest hail stone must be ¾-inch or larger in diameter. Wind gusts of 50 knots (58 mph) or greater is also a criterion for the issuance of a severe thunderstorm warning. Either the hail or the wind or both criteria would justify the issuance of this warning.

In the rare long-lasting hailstorms, storms that continue for more than twenty minutes and sometimes for over an hour, hail accumulates on the ground. Hailstorms have occurred with hail depths reaching one foot (30.5 centimeters).

Even more impressive are haildrifts. Haildrifts are formed by flows of water rushing through streets or over land, pushing the hail that has been falling into mounds of hailstones. Some flash flooding thunderstorms originate as severe thunderstorms, so that sometimes a fall of accumulating hail will be followed by very heavy rainfall, so that subsequently the runoff picks up and/or shoves the hailstones along, ultimately producing some piles of hail called haildrifts. Haildrifts of several feet sometimes occur, and haildrifts have been formed as deep as ten to some fifteen feet (about 3 to 4½ meters deep). When these moving waters hit obstacles such as buildings, the hailstones are forced to pile up along the buildings.

There are documented cases of large hail killing people, and cases where hail has killed horses and smaller animals, along with pulverizing crops, including fields of mature corn.

It is interesting to read the treatise on weather written by Aristotle. By way of some background on him:

Around 450 B.C., the Greek philosopher Socrates promoted education about everything, including Nature. One of his pupils, Plato, founded a school called The Academy in the year 386 B.C. He lectured and wrote, continuing the legacy of Socrates. One of Plato's pupils, Aristotle, who became the personal tutor of Alexander the Great, authored books and treatises in areas such as ethics and morals, politics, the scientific research method, and the natural sciences. Around 350 B.C., Aristotle wrote a book which he entitled, "**METEOROLOGICA**", from the Greek word, "meteoron", and its plural, "meteora", meaning "things in the air". The word meteor (and meteorite when it hits a planet's or moon's surface), comes from this root. Also, the word METEOROLOGY comes from the same root. Aristotle's book was the first major treatise about weather.

Here is what Aristotle wrote, so many centuries ago, about hail:

"Hail is ice, and water freezes in winter; yet, hailstorms occur chiefly in spring and autumn and less often in the late summer, but rarely in winter , and then only when the cold is less intense...Some think that...the cloud is thrust up into the upper atmosphere...and upon its arrival there, the water freezes...But the fact is that hail does not occur at all at a great height...they froze close to the earth, for those that fall far and worn away by the length of their fall and become round and smaller in size."

Thus, even in the first known comprehensive book on weather, Aristotle realized that the cause of the formation of hail is different from what causes the snow, sleet and freezing rain of wintry precipitation.

Now let us look at some photographic evidence of what hail can do.

Figure 15-9. Hail is a hazard to aviation, as depicted by this example of an aircraft that spent only about thirty seconds flying through large hail in a thunderstorm. The windshield was also shattered, but had just been replaced when this photograph was taken.

Figure 15-10. (left) Hail damage to a shingle roof near Weatherford, Oklahoma.
(source: DOA)

Figure 15-11. Hail damage to a greenhouse in the Pittsburgh, Pennsylvania area.
(source: DOA)

Figure 15-12. A haildrift in Omaha, Nebraska caused by heavy runoff from heavy rainfall pushing accumulated hail into drifts.

Figure 15-13. Hail covers the ground in a park in Denver, Colorado. Eastern Colorado and southeastern Wyoming receive some of the most vicious hailstorms. (source: DOA)

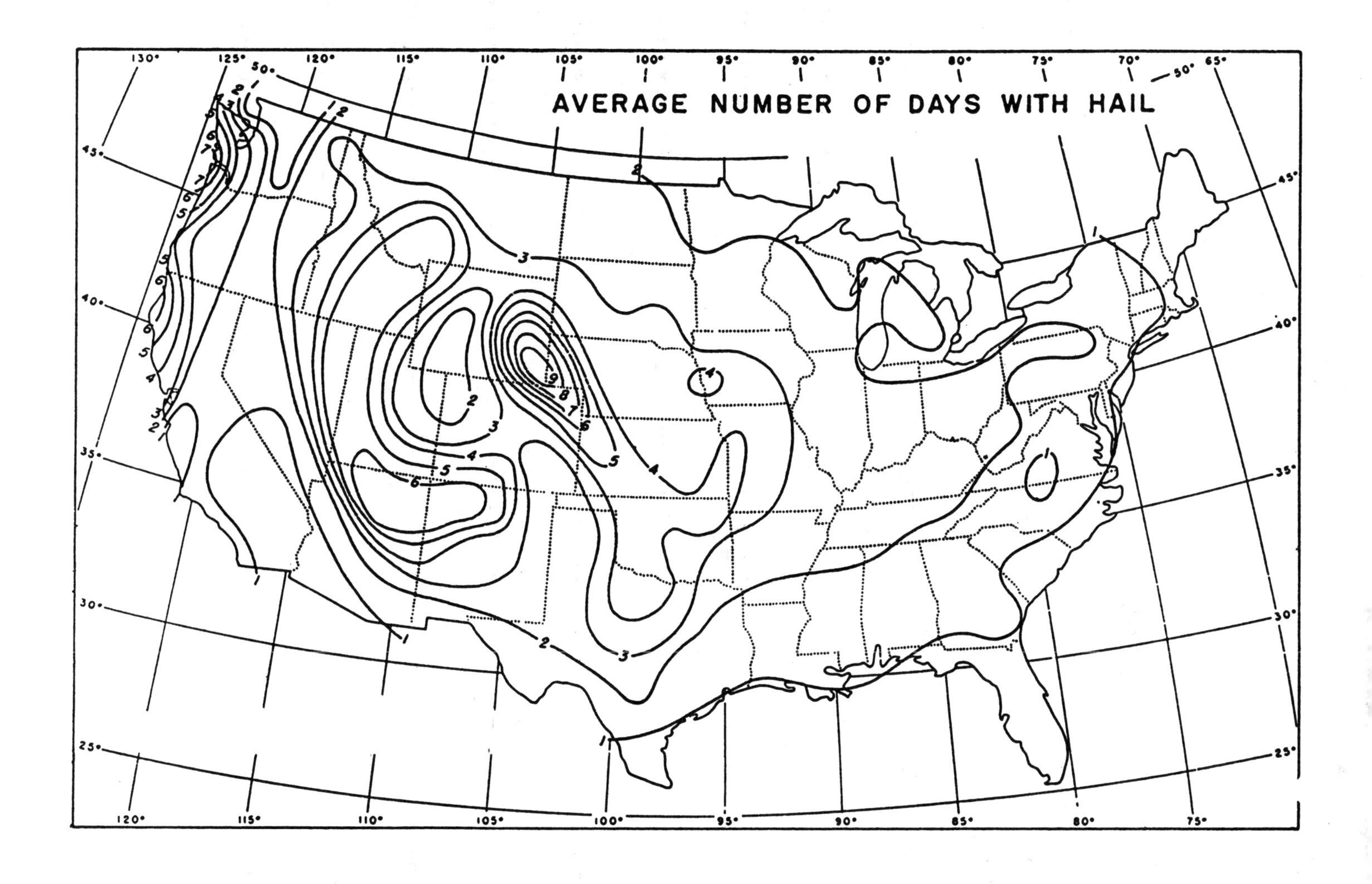

Figure 15-14. The annual average number of days with hail in the lower-48 United States. (source: DOA)

Figure 16-1. The tornado is a rapidly rotating vortex of air, having a diameter from a few hundred feet to sometimes over a mile; the air within the tornado is also rising into the cumulonimbus cloud from which the tornado forms. (source: NOAA)

Although the tornado is one of the smallest storms in Nature, it is the most violent. Winds vary from about 50 to 60 mph in the weakest "twisters" to over 300 mph in the most extreme.

They have occurred on every continent except Antarctica. The United States experiences more tornadoes than any other country, with about 1000 tornadoes reported annually. Every state of the U.S. has reported tornadoes. They occur in China and India with some frequency also, and on occasion have occurred in Europe.

Characteristics of tornadoes:

Most tornadoes come from a type of thunderstorm known as the supercell. Whereas a typical or routine thunderstorm cell has a life cycle of from 20 to 60 minutes , a supercell may persist for 3 hours or more. Most tornadoes typically emerge from the southwest part of the storm from wall clouds that are attached to a rain-free area called the rain-free base, which often follows large hail (one-inch or larger in diameter), which, in turn, follows heavy rain. This sequence of events occurs as the one looks to the southwest (first comes the rain, then the large hail, then the tornado).

When the speed of the wind increases with height, and when the wind direction profile is favorable, tornadoes are possible. A favorable wind direction profile is as follows. Suppose the surface wind is from the south, but about 1000 feet off the ground the wind is from the south-southwest, and then another 1000 feet higher the wind is from the southwest, and 1000 feet above that the wind is from the west-southwest. You can see that the wind is veering with height in the lowest part of the atmosphere. Since air is rising into the thunderstorm and is accelerating upward (an updraft), this veering wind profile also causes the air to begin rotating or swirling as it rises. The result is sometimes a mesocyclone, or a small counterclockwise rotation within the thunderstorm. (On rare occasion, the rotation can be the opposite, clockwise.) This is the incipient stage of a potential tornado.

Typically, what follows is a sudden lowering of part of the base of the cloud, which is called a wall cloud, and if this wall cloud is rotating (the mesocyclone), then the tornado formation is trying to get underway. Out of the wall cloud descends the tornado.

Sometimes more than one funnel protrudes from the thunderstorm. If a funnel stays aloft (not making contact with the ground), it is termed a funnel cloud.

Tornadic thunderstorms often produce a cloudform in another part of the storm, called mammato-cumulus (also referred to as cumulonimbus mammatus). These are pouches or cloud sections that look like huge grapes hanging from the sky. The bases of these clouds are fairly high and they do not produce tornadoes; however, they indicate that severe turbulence with powerful updrafts and downdrafts are occurring with this thunderstorm.

The word TORNADO comes from the Spanish "tornar", which means "to turn" and from the Spanish "tronada", which means "thunderstorm". A tornado is also called a twister. The appearance of the tornado ranges from slim and rope-like to elephant-trunk-like. The thinner, snake-like appearance occurs usually as the tornado is dying. In the United States, the largest tornadoes tend to occur between the Rocky Mountains and the Appalachians. For example, tornadoes in Denver are usually thinner than tornadoes in Kansas City. Tornadoes generate the best in the flatter plains areas.

Figure 16-2. A rotating wall cloud. When a part of the thunderstorm cloud base, typically in the southwest part of the storm, lowers, then that lowered cloud part is called a wall cloud. If the wall cloud starts to rotate, usually counterclockwise in the Northern Hemisphere, then a tornado may develop as an outgrowth of that rotation. (source: NOAA)

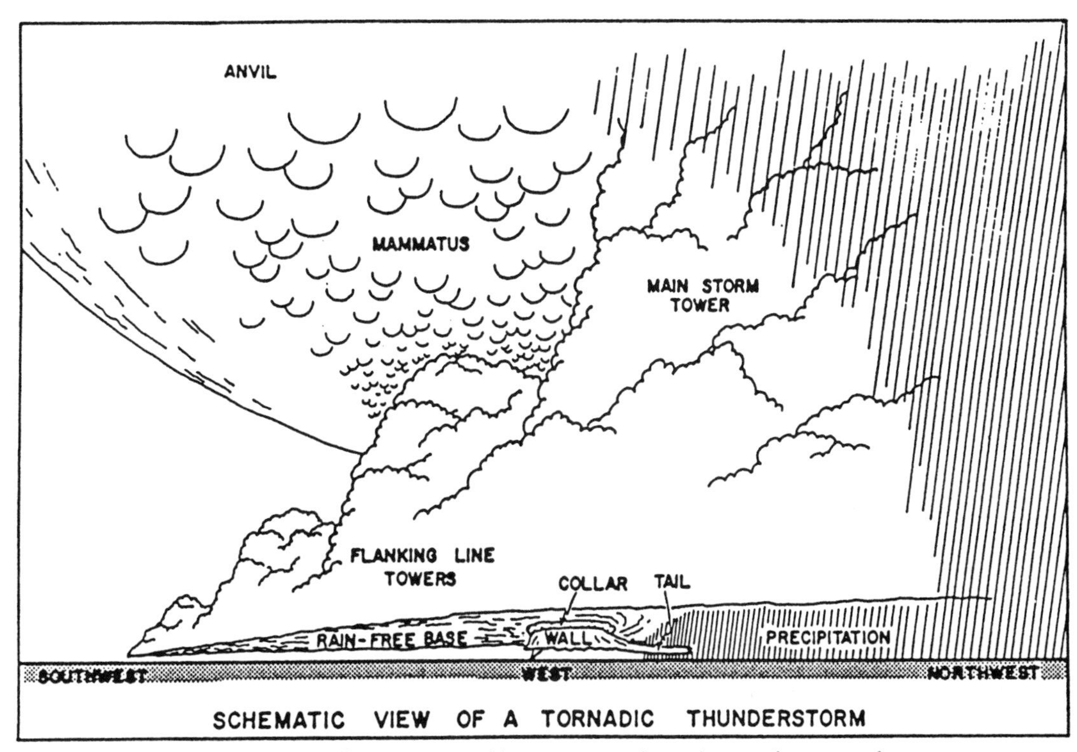

Side view of a supercell storm. View is to the northwest. Prominent features of the storm are indicated.

Figure 16-3. A depiction of what most tornado-producing thunderstorms look like. This shows how the tornadic thunderstorm would look if you were looking at it when located east of the storm, looking westward. The main updraft of warm air is entering the storm at the cloud base below the main storm tower. Strong winds aloft are blowing from the southwest to the northeast. Air in the upper portion of the updraft eventually becomes colder than the surrounding air. At this level, the cloud spreads out rapidly, forming an anvil. The term "wall" stands for wall cloud, "collar" is for collar cloud, which is a circular ring of cloud surrounding the upper portion of the wall cloud, and "tail" is for tail cloud, which is a low tail-shaped cloud extending outward from the northern quadrant of a wall cloud. "Flanking line towers " refers to a line of cumulus/towering cumulus clouds connected to and extending outward from the most active portion of the parent cumulonimbus. (source: NOAA)

Tornadoes do fill in with cloud matter, but also take on the color of the debris they are carrying. For example, when a tornado passes over a field of plowed topsoil, it will lift some of the topsoil to appear black. A tornado in Nebraska passed over heavy snow cover and became a white tornado, filled with snow, which is unusual.

Sometimes the tornado vortex is initially invisible, but the observer will recognize that a tornado is occurring because debris will begin swirling around at the tornado's base and will eventually be lifted into the funnel, making the funnel visible. Moreover, the funnel soon fills will cloud material, making it visible.

Sometimes the tornado cannot be seen because of poor visibility due to flying debris, heavy rain and low clouds. At night, illumination by lightning should show the vortex. Although there is typically no precipitation occurring where the tornado is located, some tornadoes in the southeastern United States are enshrouded in heavy rain, making their visible detection difficult, especially at night.

Hurricanes making landfall can also spawn tornadoes. When a hurricane moves inland, the frictional action of the land surfaces, plus the hurricane losing contact with its chief source of energy, the warm moist oceanic air, weaken the wind at the surface. But the wind a little higher, at about 2000 to 4000 feet off the ground, remains at full fury for a hours longer, before it is eventually lowered. So what is happening is that the wind shear dramatically increases in the first few thousand feet, creating a favorable environment in the warm,moist unstable air, for tornadoes. Hurricane tornadoes tend to occur in bunches and are usually smaller and shorter-lived than supercell thunderstorm tornadoes.

Even if you cannot see the tornado, you should be able to hear it. The strongest tornadoes, with their furious wind speeds, sound somewhat like an approaching freight train. Their wind speeds range from about 40 to 50 mph (about 65 to 80 km/hr) in the weakest or minimal tornadoes, to over 300 mph (over 480 km/hr) in the most powerful. Tornado intensity is rated on a Tornado Force Scale developed by Ted Fujita, wherein F0 (F-zero) is a minimal tornado, F1 is stronger, etc., until F5 for is reached for the strongest tornadoes. Most tornadoes are F0s through F3s.

Tornadoes range from a few hundred feet diameter to over one-half mile. Sometimes tornadoes over a mile in diameter occur. The biggest tornadoes typically last the longest. Tornadoes may last for from a few minutes to over a half hour. Supertornadoes can persist for up to several hours. Most tornadoes move at about 30 to 45 miles per hour (about 48 to 72 kilometers per hour), although some have been observed to remain stationary for a while, and some have been clocked to move at around 70 mph.

Figure 16-4. Most tornadoes last from several minutes to about a half-hour. Huge tornadoes can persist for over a hour. They typically move at from 20 to over 50 mph with winds inside the funnel of from around 50 mph to as extreme as over 300 mph! Their widths range from a few hundred feet to sometimes over a mile. The monster ones can pick up people and automobiles and fling them through the air. (source: NOAA)

Since the winds are so powerful in such a small storm, and since wind is caused by air blowing from high pressure to low pressure, we can therefore estimate what the pressure gradient must be from the outside edge of a tornado to its center. If we know the radius of the tornado, we can estimate the central pressure. In the F2 through F5 tornadoes, the air pressure in most of them is probably in the range of 15" to 22" on the barometer. Recall that average sea-level pressure is near 30", and the height of half the atmosphere would be where the barometer reads 15", the 500 millibar level, which is about 18,000 feet up. Thus, in the bigger twisters, the pressure is what you would expect to find at from about the 730 millibar level to about the 500 millibar level.

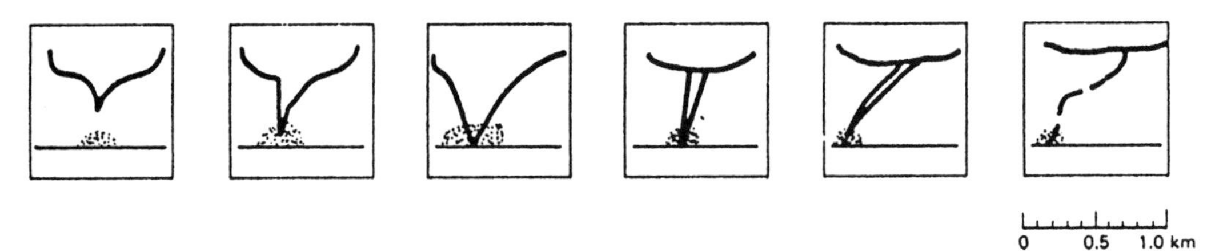

Figure 16-5. The life-cycle of a typical tornado. (source: Environment Canada)

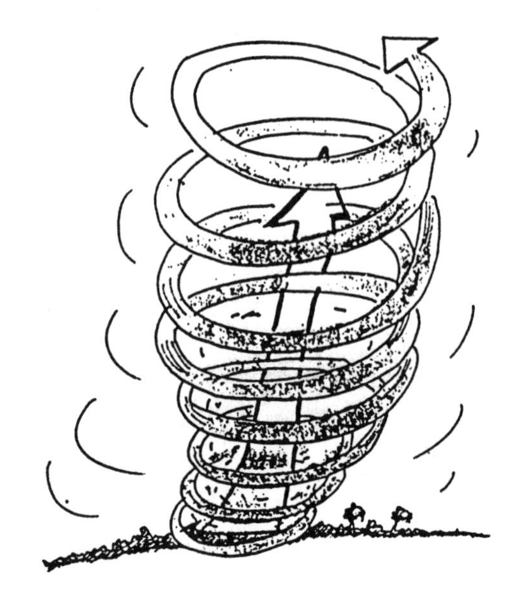

Figure 16-6. The air is swirling as well as rising around the center of the tornado. Tornado-genesis requires the right proportion of buoyancy of the local environment and wind shear, especially shear in wind direction. Here is the explanation of these conditions. Buoyancy of the local environment means that air parcels are warmer than the environment, so they become buoyant and rise. The wind shear is the change of wind direction or speed with height. For tornadoes, we are looking at the wind shear in the first few thousand feet from the ground up. Let us say, e.g., that the wind at the surface is from the south at 15 mph, and at 1000 feet up the wind is south-southwest at 20 mph, and at 2000 feet up it is southwest at 25 mph, and at 4000 feet up it is west-southwest at 35 mph. You can see that rising air parcels, called updrafts, would be twisting as they rise. If they accelerate, they could form a localized rotation in the thunderstorm cloud, leading to tornado development. If the air is too buoyant, then whole layers of air rise, which is too difficult for the local wind shear to start rotating to form a tornado. Thus, the ratio of buoyancy to shear must lie within a specific range to permit the development of tornadoes. (source: NOAA)

Destructive agents:

A major tornado destroys a house by the combined action of intense winds and suddenly much lower pressure. By relating the intense wind speed to the size of the funnel, we can estimate the pressure-gradient force in the tornado. This force implies that the barometric reading inside a major tornado is under 20 inches, perhaps in extreme cases under 15". This is typical of air pressures found at some 10,000 to 20,000 feet high. Such drastic pressure changes that are almost instantaneous can cause enclosed structures to explode as the inside air pushes outward. The extremely low pressure is caused by the rapid outflow of air from the top of the storm, and is intensified by the failure of the incoming air to make up this loss.

When a tornado passes over a home, the roof usually comes off first and then the destructive winds continue to do more damage. The reason the roof usually comes off initially is because most homes have roofs that overlap the sides of the house, and since the air in a tornado is not only swirling around but is also rising, this rising air lifts the roof off. Thus, a house's roof becomes like an airplane wing. Having many heavy-duty leather or metal straps to fasten the roof to the house would be helpful in tornado-prone areas; this could be done during the initial home construction so that these tornado fasteners are inserted on the inside of the house and would be hidden by the halls and ceilings. These strips are also called hurricane strips since they serve a similar function in areas prone to being struck by hurricanes.

As the winds of the tornado pick up debris and fling it through the air at high speeds, the debris become missiles; many of the tornado deaths result from fast-moving debris striking people. A classic example of the effect of tornado winds is that of a straw driven into a telephone pole.

To help prevent flying objects from crashing through glass windows, some public and private buildings in tornado-prone areas use laminated glass, which is similar to the glass used in automobile windshields. This type of glass is comprised of at least two pieces of glass with a tough plastic liner placed between them. The result upon impact is that the glass typically will not shatter when it breaks.

When a meteorologist investigates storm damage from a suspected tornado, how the damage is distributed is a key to identifying whether the storm that produced the damage was a tornado. Tornado damage is typically flung around in a circular/oval fashion, whereas straight-line wind damage is typically blown over in one direction.

How far up into the cumulonimbus thunderstorm cloud does a tornado extend? Based on research from Doppler weather radars, which show the wind profile in the clouds, the answer is that the vertical extent of the tornado in the thunderstorm varies. Small tornadoes may extend upward for thousands of feet, and the strongest and largest tornadoes may extend upward past the middle of the thunderstorm. For example, if a particular thunderstorm has clouds that grow to 50,000 feet above the ground, then the largest of tornadoes may extend from the ground to some 35,000 feet up.

Places and times of occurrence:

Because of the geography and climate of the United States east of the Rocky Mountains, this region is the most tornado-prone in the world. Tornadoes have occurred in all 50 of the United States, but the most vulnerable regions are the area between the Rocky Mountains and the Appalachian Mountains, and the Florida peninsula. Oklahoma is at the strongest risk for having tornadoes, especially the central part of that state. From the figure below, you can see that the region extending from northern Texas through Oklahoma into Kansas and Nebraska is the main "TORNADO ALLEY". The peak tornado season starts early in the year in the southeastern states and gradually moves northward during the spring, peaking in May in the central plains, and then moves northward into the northern plains during the summer when it peaks there. Most tornadoes occur during the afternoon and evening.

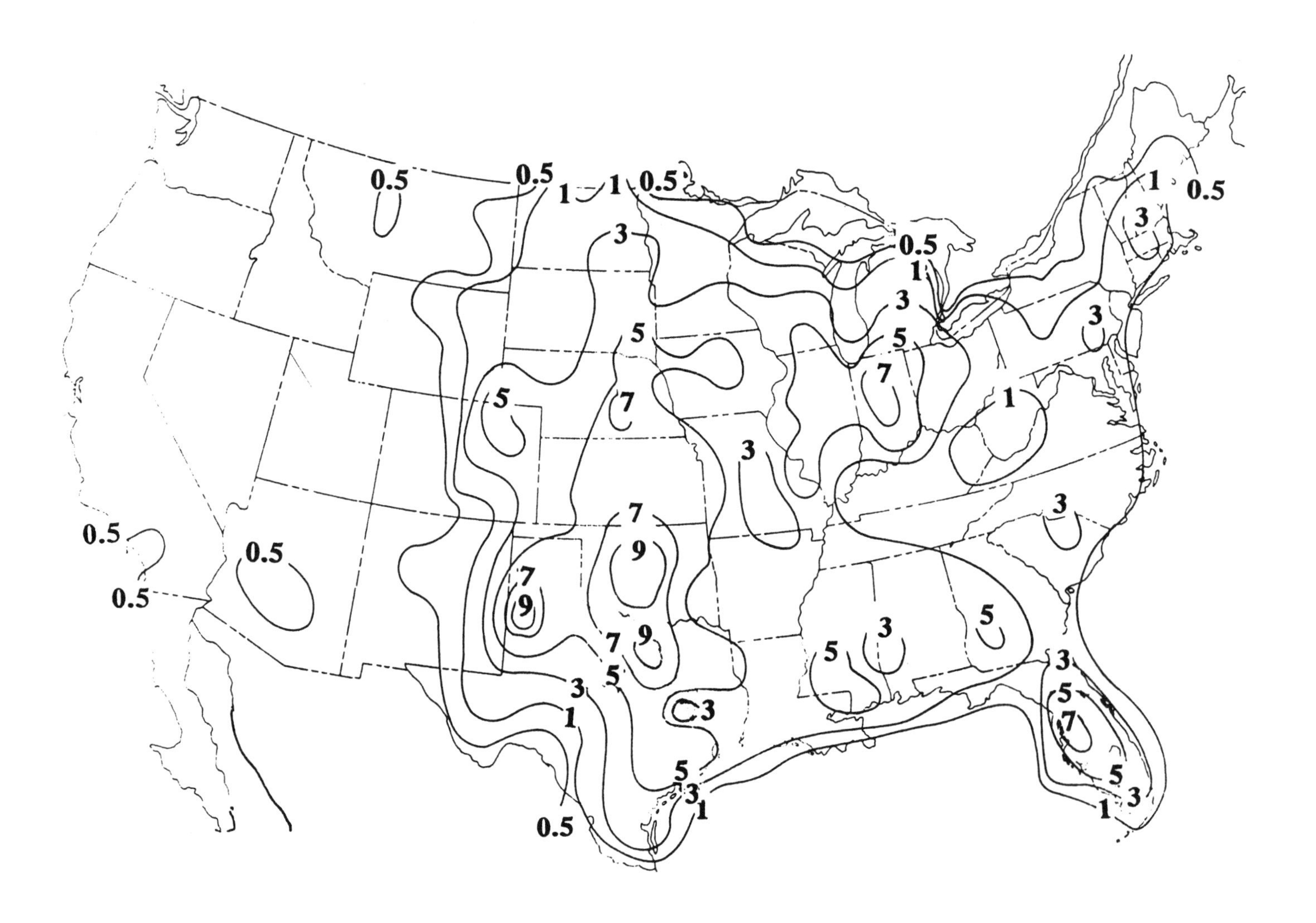

Figure 16-7. The number of tornadoes annually per area of 100 miles long by 100 miles wide. Note that the highest figures are in north Texas, central Oklahoma, south-central Kansas, eastern Nebraska, central Indiana and the Florida peninsula. (source: NOAA)

Figure 16-8. The Tornado Force Scale, developed by T. Fujita.

F0	Gale Tornado	weak	40-72 mph
F1	Moderate Tornado	weak	73–112 mph
F2	Significant Tornado	strong	113–157 mph
F3	Severe Tornado	strong	158–206 mph
F4	Devastating Tornado	violent	207–260 mph
F5	Incredible Tornado	violent	261–318 mph

Estimate of F-scale Wind from Structure Type and Damage Category

Structure Types		DAMAGE CATEGORIES						
		No Damage	Minor Damage	Roofing Blown off	Whole Roof Blown off	Some Walls Standing	Flattened to Ground	Blown off Foundation
Outbuilding Mobile Home	□	F 0	F 0	F 0	F 1	F 1	F 1	F 2+
Weak Frame House	▱	F 0	F 0	F 1	F 1	F 2	F 2	F 3+
Strong Frame House	□	F 0	F 0	F 1	F 2	F 3	F 4	F 5+
Brick Building	⊠	F 0	F 1	F 2	F 3	F 4	F 5+	F 5++
Concrete Building	■	F 1	F 2	F 3	F 4	F 5+	F 5++	F 5+++

MINIMUM WINDSPEEDS: F 0(40mph) F 1(73) F 2(113) F 3(158) F 4(207) F5(261mph)

Waterspouts:

Figure 16-9. A waterspout. Tornadoes can move across rivers, ponds and small lakes. They are still tornadoes. Sometimes, funnels form over large lakes or over the ocean and look like thin tornadoes but are much weaker than most tornadoes (they would be an F0 or F1 on the Tornado Force Scale); such funnels are called waterspouts. When they do move onshore, they may cause minimal damage. (source: NOAA)

Cold-air funnels:
During cool weather in northern states, such as around Lake Ontario in the autumn, funnels may form which look like thin tornadoes. These are cold air funnels and are not particularly dangerous.

Although tornado season is primarily spring and early summer, tornadoes do occur in the winter, primarily in the southeastern United States. These winter tornadoes, unlike the typical tornado, can be enshrouded in rain, yet still be major tornadoes.

Watches and warnings:

When the atmospheric conditions are prime for tornado development, or are expected to become ideal, then the National Weather Service issues a tornado watch. When a tornado is developing, or when imminent development is suspected, or when a tornado is already in progress, then the National Weather Service office that has forecast and warning responsibility for that area issues a tornado warning.

Local radio and television broadcast severe weather warnings, as does the weather radio. Since the warnings are issued by the National Weather Service, they will be immediately disseminated on the weather radio. Weather radios have a tone alert feature, so that as the warning goes out, even if it is the middle of the night, a loud tone at the 1050 Hertz frequency will alert you to the potential danger. Many stores that sell electronics also market weather radios.

Shelter from a tornado:

If caught in an open area outdoors, try to determine which way the tornado is moving. In the Northern Hemisphere, tornadoes move most often towards the northeast. Run at right angles to its path. If the tornado is going to pass over or very near you, or when its debris is being hurled your way, jump into a ditch. If on a highway, get under the girders of an overpass. A strong tornado can pick up and hurl an automobile with people inside through the air.

In a building, go the basement and take shelter under a heavy piece of furniture. If there is no basement, get away from windows and move to the interior of the house. A door opening of a closet is also recommended. Cover your head first and try to cover your entire body to protect it from flying missiles and other debris. If time permits, you can pull a mattress over yourself.

Trained weather spotters:

In the United States, the National Weather Service has the responsibility for issuing tornado watches and tornado warnings. A tornado watch means that the weather conditions are favorable for tornado development; a tornado warning means a tornado is developing or is already underway. Persons in the path of the tornado must take immediate shelter to save themselves from injury or death. As part of its mission, the National Weather Service trains volunteers who watch the skies for tornado development. These persons are called **weather spotters (Figure 16-10, above)**. Their reports along with data from the velocity display of Doppler weather radar are responsible for information that leads to tornado warnings that routinely save large numbers of lives. Local law enforcement personnel and others are also often trained in weather spotting for severe weather. (source: NOAA)

Some tornado extremes: The widest tornadoes are over a mile in diameter. The fastest moving have been known to race forward at up to about 70 mph (near 115 km/hr), and the longest lasting tornadoes have stayed on the ground for over 100 miles, even for more than 200 miles, but this is quite rare. On April 3rd-4th, 1974, a major tornado outbreak resulted in about 150 tornadoes touching down in eleven U.S. states and in Ontario Province, Canada. And in June 1995 at Lazbuddie, Texas, SIX tornadoes were on the ground at the same time from the same thunderstorm!

DEMONSTRATION: MAKING A TORNADO MODEL

Introduction:

We will attempt to make a model that will imitate some of the conditions necessary to form a tornado. We will need an unstable air mass; that is, we need one in which the air is very warm at the bottom and relatively cool at top. This is achieved by heating the air from below with the resulting convection increased by having the heated air rise through a chimney. The rising air must have a rotating motion. This is done by allowing the cool air that enters the model to force the hot air up the chimney. Steam is condensed into water droplets to make the rotating warm air visible as a cloud-funnel does in nature.

Materials needed:
(for non-U.S. readers, use 1 inch = 2.54 centimeters, or 1 cm = 0.394")

- 4 sheets of masonite, 7½" x 8"
- 1 sheet of masonite, 8" x 8"
- 1 metal baking pan, 8" x 8" x 2"
- 2 sheets of clear plastic, 6" x 6½"
- 1 stove pipe, 30" to 36" long, with a 3" diameter
- 4 pieces of stiff wire, each 2½" long (you can use wire from opened up paper clips)
- 4 strips of wood (molding), each about ¾" x ¾" x 6"
- flat black paint
- waterproof glue
- short screws
- masking tape
- a light source, such as a slide projector
- a heat source, such as a stove or hot plate

Procedure for constructing the model:

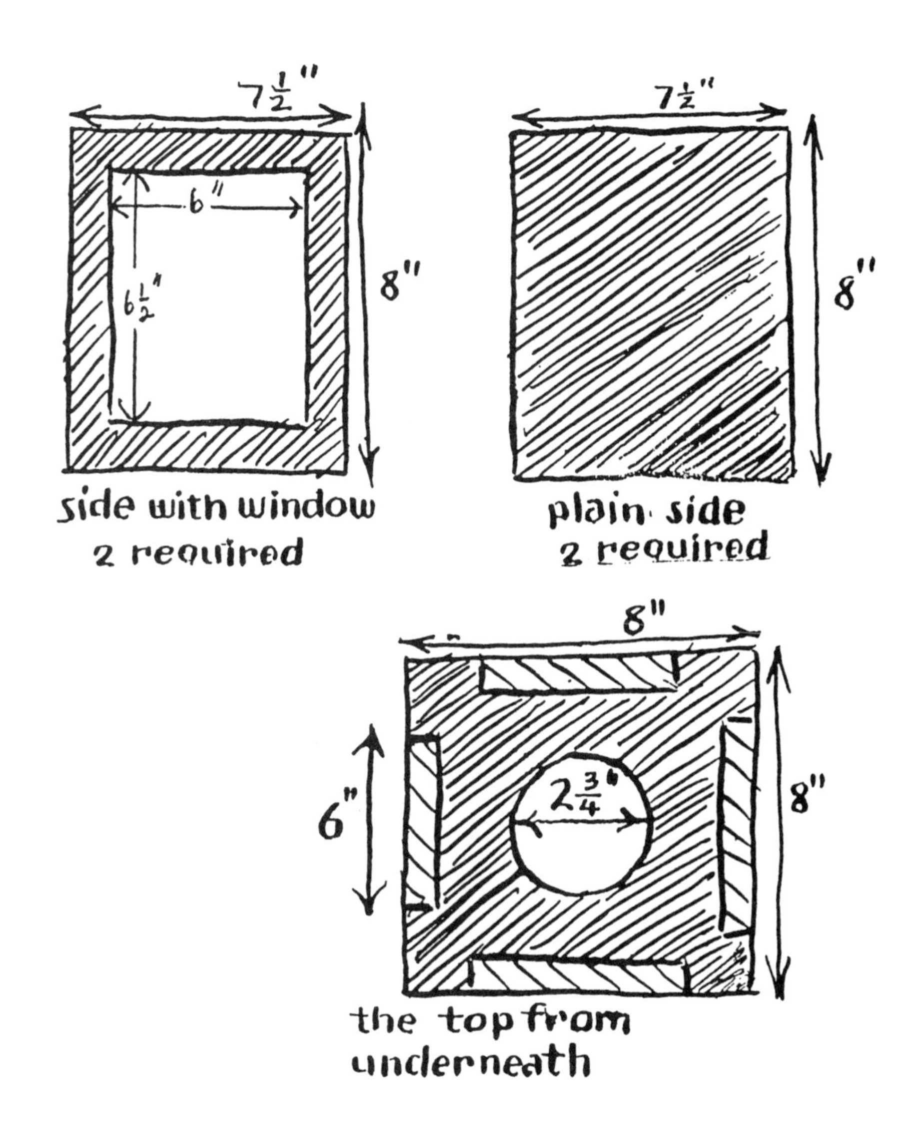

1. Cut out a square of about 6" x 6½" from two pieces of 7½" x 8" sheets of masonite. Leave the other two pieces alone for now. These will make up the sides of the box, which will fit into the baking pan later on. See the sketches above.
2. Carve a circle with a diameter of 2¾" out of the center of the fifth sheet of masonite, which is 8" x 8", and will be the top of the box. See the diagram above.
3. Screw four strips of wood molding of about ¾" x ¾" x 6" to the underside of the top sheet. This is where the sides of the box will be attached to the top.
4. Paint the surfaces that will be the inside of the box with flat black paint.
5. When the surfaces are dry, fasten the sides of the box with small screws to the top piece so that the two windows are next to each other and there is a half-inch slot on the right-hand side of each of the four sides as shown in figure 32-11 which appears at the end of this demonstration instructions.
(continued on next page)

6. About (3/8)" above the bottom edge of each side, punch or drill a small hole about 1" from the left end and another hole about (3/8)" from the right end. See the diagram after step 7.
7. Cut four pieces of stiff wire, each about 2½" long, from the opened paper clips or other source, and pull them through the holes across each corner. See the diagram below.

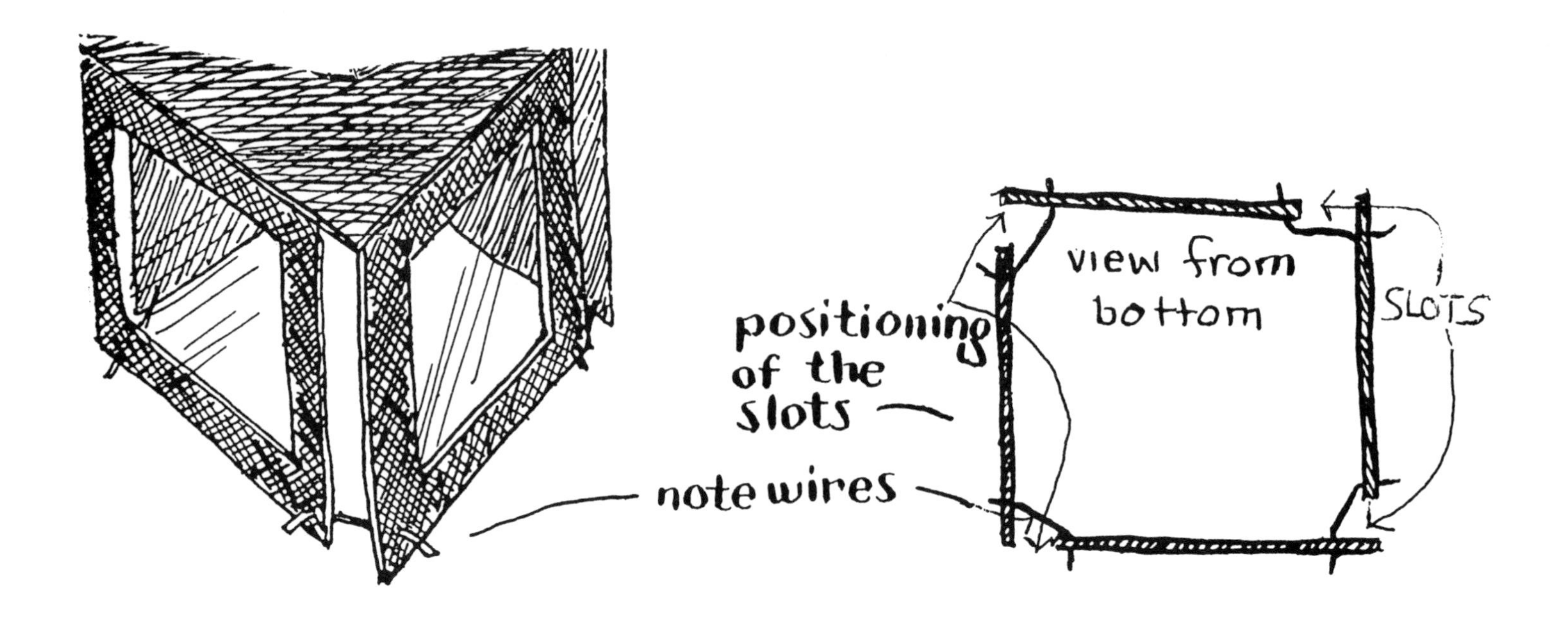

8. Set the box on the pan so that the rim of the pan is inside the box, and the wires rest on the pan's corners. Press the wire ends tightly against the box sides and pull them while bending, so that the box sits fast on the pan.
9. Glue or tape the plastic sheets over the cut-out area of the two window frames.
10. Fasten the stove pipe to the top of the box with masking tape. The diameter of the circle cut out of the top of the box is slightly smaller than the diameter of the stove pipe, in order to assure a tight fit.
11. Make sure that there are no air leaks between the chimney and the box, between the sides and top of the box, and between the box and pan. Air should enter the box only through the slots.
12. Feel free to make any improving modifications to this model.

Using the model to simulate a tornado:

1. Fill the pan through a slot with hot water to within one-half inch from the rim.
2. Place the entire model on a stove or hot plate.
3. The water should be heated until steam rises and condenses to tiny droplets of water above the pan. The water should approach the boiling point but not boil vigorously.
4. Shine a bright light source, such as from a slide projector, through one of the windows and observe through the other. Wait a few minutes. What do you see?
5. The motion of the water can be made better visible by blowing a small amount of chalk dust into the water.
6. The funnel cloud can be made more visible by blowing smoke into the box.
7. What would happen if the half-inch slots were on the left-hand side instead of being on the right-hand side?

This demonstration is a modification of the one described on pages 104 to 107 of the Manual of Lecture Demonstrations, Laboratory Experiments, and Observational Equipment for Teaching Elementary Meteorology in Schools and Colleges by Hans Neuberger and George Nicholas of The Pennsylvania State University. **Figure 16-11, below, shows this device.**

DEMONSTRATION: A SIMPLER VERSION OF A TORNADO MODEL

Materials needed:
2 transparent two-liter soda containers from which the labels have been removed
3 inches of plastic tubing that will fit over the opening of each container. Usually, tubing with an internal diameter of 1 inch will fit over most containers; however, if you want to be sure, take your two empty containers to a local hardware store and select the tubing that best fits your containers.
blue food coloring

Procedure:
1. Place about 3 inches of plastic tubing over the opening of one of the 2-liter containers. The tubing should fit snugly. Now fill the container with water.
2. Invert the other, empty, container and attach its opening to the other end of the plastic tubing.
3. Invert the container that is filled with water over the empty container. What happens?
4. Now place the bottom container so that it is on top of the empty container, spinning it in a clockwise direction as you do so. What happens?
5. Repeat step 4 while spinning the top container in a counterclockwise direction.
6. If you want the water to appear more prominently in the containers, then add several drops of blue food coloring to the water.

DEMONSTRATION: CREATING A TORNADO IN A ROOM, FROM FLOOR TO CEILING

You can do this demonstration if you have access to a room that has an exhaust ceiling fan. You would need to also suspend about 3 room fans from the ceiling around the ceiling fan so that you can create rotation of the air as it rises. If you can do this, then get some "dry ice" (solid, i.e., frozen, carbon dioxide) and place it on a piece of cardboard directly under the ceiling fan. Never touch the extremely cold "dry ice" with your bare flesh!!! Start all the fans. Have long black sheets taped on the wall as a background so that you can see the funnel better when you look at it with the black background behind it. The dry ice will cause water vapor in the room to condense, forming a white cloud, which rotates as a vortex, stretching from the floor to the ceiling. It is quite impressive! You can even put miniature figurines of people, animals and cars into your tornado.

Now let us change the focus to a whirlwind that is not a tornado but is like a little cousin of the tornado: the dust devil.

Dust devils (sometimes also called whirlwinds and dust whirls):

Dust devils are swirls or vortices of air that look like little tornadoes, but whose winds are typically from 35 to 60 miles per hour, rarely somewhat higher. They are caused by very localized intense heating of the surface, so they occur during the greatest heat of the daytime. If you drive in Nevada from Las Vegas to Reno on a sunny and very hot summer day, you often see dust devils in the desert. If you run into one, you will be sandblasted by winds of about 50 miles per hour or somewhat more. They can also form on a sunny cool day when intense local heating of the surface creates an unstable column of air and creates dust devils that can rotate either counterclockwise or clockwise, filling with dust, leaves and small pieces of debris such as paper.

It is unusual for this type of whirlwind to cause serious trouble, although some cases have been investigated and documented that were more serious. Consider, for example, these following cases from New England.

In Dudley, Mass., a 3-year old child was picked up and tossed about ten feet, giving her a cut and some bruises.

In Ipswich, Mass., a 9-year-old girl and her bicycle were picked up by an unusually strong dust devil, and were flung about 40 feet across the road. The girl's shoes were carried about 100 feet. She, too, was cut and bruised, but not seriously. She described the sound as "awful, like the roar of a lion".

Near Northboro, Mass., eight turkey feeders, some weighing about 300 pounds, were toppled, and their covers, which were 3 x 10 feet, were sent spinning through the air. The dust devil also lifted a hood from a truck, pushed a calf into a fence and broke a large tree branch before disappearing. A witness described the sound as that of a baritone siren of a freight train.

In Methuen, Mass., a dust devil damaged a roof of a home and picked up some lawn furniture.

In Belfast, Maine, a dust devil picked up a garage, spun it and carried it nearly 100 feet, demolishing it. The dust devil moved southeastward about one-half mile until it reached the bay. Then over water it picked up a column of spray to a height of 12 to 14 feet, looking like a little waterspout, and then continued about another half-mile before dissipating.

Thus, although dust devils are usually relatively harmless, there are rare events of their being stronger than usual.

The Coriolis force due to the earth's rotating is negligible in such vortices, since the rotation is occurring over a very small area. The wind in a tornado, around the eye of a hurricane and in a dust devil is a balance between the pressure-gradient force and the centrifugal force; such wind is called a cyclostrophic wind.

Now, here are some interesting tornado stories:

On average, the lower 48 states experience about one thousand tornadoes annually. Some years have considerably fewer while some years have major outbreaks. It is likely that some small and short-lived tornadoes in open country are never reported.

Big tornadoes are easy to detect, but the small ones are now likely to be more reported because of excellent weather spotter networks established by the National Weather Service, television and radio meteorologists, amateur radio operators and other groups, and also because the Doppler weather radar better depicts tornado signatures than did the former generation of weather radar.

With the proliferation of personal video-cameras, we now have many excellent close-up videotapes of tornadoes in action. We know that besides the winds rotating with speeds of from near 100 mph in the weakest tornadoes to well in excess of 250 mph (even over 300 mph) in the most intense, the air inside a tornado also rises. Thus, when a tornado hits the house, the roof is usually the first part of the house to be ripped off because a roof typically overhangs the sides of the house, causing it is act somewhat like the wings of an airplane when the rising air inside a tornado passes over the house. Thus, the roof is lifted off, leaving the inside exposed to the destructive winds.

An engineering study of buildings demolished by tornadoes suggests that if "hurricane straps" are used to help secure the roof to the frame, then the structure may have a better chance of surviving at least small and moderate tornadoes. Hurricane straps are leather or metal strips with holes in them, so that one end attaches to the bottom of the roof and the other end to the top of the framing. These inexpensive straps are placed around the frame of the house at intervals of a foot to eighteen inches. Extra building support is attempted by building the structure with additional two-by-sixes or two-by-eights, e.g., in the corners of the building.

The effects of the wind and drastic pressure drop when a tornado hits are well known. Moreover, tornadoes sometimes cause results which are worth documenting because of their uniqueness or because these results are unusual or intriguing.

For example, in May of 1957 a tremendous tornado smashed through what is now the southern part of Kansas City, Missouri, destroying homes by the hundreds and killing people as it carved out a path up to a half-mile wide. A surprise occurred when the tornado hit the high school in the community known as Ruskin Heights.

The following story was told to this author by a teacher who had worked at that school at that time.

A sign across the front of the school read, "RUSKIN HEIGHTS HIGH SCHOOL". When the tornado hit the building, it destroyed much of it, but much of the front of the structure was still standing. The tornado blew all the letters off the sign except for the letters
R U I N.

Nature can often produce stories better than those we can create from our imaginations.

Loss of life also occurs because of objects that become flying missiles. One tornado watcher narrowly escaped death while he was taking photographs of an approaching tornado. A white object was being rotated in the twister as the funnel approached him. The object became bigger and bigger to him as the tornado got closer. Just before he had to stop taking photos and run for cover, he saw that the object was a pick-up truck that the tornado had "picked up" and was violently whirling around.

The most intense tornadoes can pick up automobiles, people, livestock, and derail locomotives. On occasion, a person or a dog is picked up by a tornado, carried some distance and gently deposited down on the ground or on top of a tree.

Tornadoes have also been known to drive straws into telephone poles.

One strange tornado tale is of a twister that passed over a marshy region, scooping up frogs and tadpoles. The funnel retreated into the thunderstorm clouds, and when those clouds later passed over the next town, it dropped the frogs and tadpoles. In other words, it was raining frogs.

A woman called this writer when I was doing a radio interview talk-show in Rochester, New York, to convey the following story. She was at home in Minnesota when she suddenly heard several loud thumps on the roof of her house. Moreover, some objects were also plummeting into the ground. When she looked outside, she saw box-turtles falling from the sky. This apparently also happened with severe thunderstorms in the area.

In 1931 a Minnesota tornado lifted a 166,000-pound railroad train with 117 passengers aboard, carrying the load for some 80 feet. A Mississippi tornado in 1975 carried a home freezer for more than one mile. Another Mississippi tornado hit a house in which a mother and her daughter had time only to jump into the bathtub for shelter. When the tornado struck the house, the only room left was the bathroom. A tornado in Wichita Falls, Texas caused a similar episode to an off-duty meteorologist whose home was hit while he was inside. He grabbed a mattress and jumped into the bathtub before the storm hit. The only things left after the tornado struck were the meteorologist and his bathtub. One story is of a house demolished by a tornado, leaving only the Holy Bible left, with it opened to a page referring to the wind. Thus, meteorology has many interesting stories, and tornadoes contribute to some of the most unusual.

Figure 16-12. Mammatocumulus clouds, also called cumulonimbus mammatus. These mid-level clouds are caused by turbulent conditions in a thunderstorm which is usually severe and which can produce a tornado. The tornado never comes out of the clouds you see here. These pouches show surges of air descending with evaporation occurring as the clouds then disappear below their visible bases. Mammatocumulus clouds occur along any edge of a thunderstorm which is usually severe. When you see these clouds and they are associated with an active thunderstorm, then they serve as a warning that the storm is probably severe and capable of producing a tornado.

Life-cycle pictures of a tornado:

Figures 16-13. The lowering of the cloud base is called the wall cloud. When the wall cloud starts rotating, usually in a counterclock-wise direction, or left-to-right, in the Northern Hemisphere, it is called a mesocyclone or "meso" for short. This is dangerous since next may descend a tornado. (source: NOAA)

Sometimes you may not even see the funnel until it reaches the ground and begins to pick up debris.
Eventually, however, the column does fill with cloud matter and it is visible. The tornado funnel extends well into the cloud, and in huge tornadoes, they may extend almost to the top of the cumulonimbus cloud. (source: NOAA)

Figure 16-14. When the funnel does not extend to the ground, it is termed a "funnel cloud"; when it reaches the ground, the funnel is called a "tornado". In common vernacular, the storm is also referred to as a "twister". Most tornadoes come out of supercell thunderstorms, and move at from 15 to 50 miles per hour, though speeds have been clocked at up to some 70 mph (over 110 km/hr). Rarely, then can remain stationary for a short while. (source: NOAA)

Figure 16-15. As the tornado dissipates, it may slim to what is called a "rope stage" and then break apart, or it may just start breaking apart even if it is a wider funnel. (source: NOAA)

Figures 16-16. In the lower 48 contiguous United States and across southern Canada, tornadoes of all sizes occur from the Plains States or Plains Provinces, respectively, to the East Coast (left figure). From just east of the Rocky Mountains to the West Coast, tornadoes tend to be narrower and less violent (right figure). Tornadoes are rare from the central Rocky Mountains to the West Coast. (source: NOAA)

Figure 16-17. Occasionally, multiple tornadoes may occur from the same severe thunderstorm. Or, as in the photo at right, they may occur side-by-side as in set of twin tornadoes that occurred east of Elkhart, Indiana. Sometimes one supercell or another severe type of thunderstorm produces a series of tornadoes or one tornado that goes down and up into the clouds up to several times. And sometimes a series of tornadoes can be misinterpreted as one long-lasting, long-path tornado, although long-time ones do also occur. (source: NOAA)

Even after a tornado returns into the cloud or dies, some debris may be carried for miles before it is dropped to the surface. Thus, a case such as "raining frogs" or "raining turtles" is caused by a tornado passing over an area that has many such small animals, sucking many of them up, and then dropping them later on.

Figure 16-18. A terrifying scene: a large tornado marching towards a city. (source: NOAA)

In the Northern Hemisphere, if a tornado is about to strike your area, you typically first experience the gust front with strong, gusty winds as the storm moves in, a lot of lightning and thunder, heavy rain, then large hail in many cases, and then the tornado with no precipitation. However, where dewpoints are very high, such as in the southeast states of the United States, tornadoes sometimes are obscured in heavy rain. Most of the time, however, it is not raining and/or hailing where the tornado is occurring, and a clear sky is typically visible soon behind the tornado, since the twister's occurrence is near the edge of the thunderstorm in that southwest part of the storm.

Figure 16-19. Waterspouts. When a tornado forms over water, it is called a waterspout, and is typically of F0 or F1 strength, rarely of F2 or stronger. However, it is possible for a huge tornado, F3, F4 or F5, to move over a body of water and maintain its ferocity at least for a while. (source: NOAA)

TECHNICAL SECTION - The following chapters of this book are technical and are presented for the use of meteorologists, students and others wanting or needing such information.

Chapter 17. THE SKEW-T LOG-P THERMODYNAMIC DIAGRAM

Meteorologists use a thermodynamic diagram for many forecast uses: to forecast thunderstorms and the probable intensity, to forecast clouds, to forecast overnight low temperatures and other useful weather aspects. On the diagram is plotted the data collected by instruments attached to weather balloons, and some of the data gathered by weather satellites. The specific data is called **sounding data**, so-named because instrumentation sounds out the vertical profile of the lower atmosphere. **A sounding is a probe of the environment, in order to acquire data for scientific analysis.**

Twice daily, and sometimes more often in potentially severe weather, weather balloons, filled with helium, are released, which ascend to about 100,000 feet, expanding as they rise. A parachute is inserted into the balloon before it is released so that when the balloon bursts, its instrument package gently lands on the surface. Powered by a battery, the instruments radio back the temperature, dewpoint and air pressure continuously as the balloon ascends. The wind direction and speed at the various levels are determined by following the motion of the balloon. The instrument package is called a **radiosonde**, because it radios back the information of the sounding.

A second method for obtaining soundings is via weather satellites. Weather satellites have sensors that give us visible, infra-red, water vapor and other imagery of the tops of clouds and, where there are no clouds, of the surface of the planet. Visible pictures are useable obviously only during the daytime. Infra-red imagery senses the temperature of the tops of the clouds or the ground, water or ice surface. The Stefan-Boltzmann Law in physics relates the emitted energy of the surface of an object to the fourth power of its absolute temperature. The more radiation emitted by each square centimeter surface area of an object, the higher the object's surface temperature.

The meteorologist takes the sounding information and plots it on a **thermodynamic diagram** in order to analyze the vertical profile of the lower atmosphere over that general location. Sounding data from all sounding sites across the globe are the chief source for upper air weather maps and computer model projections of atmospheric conditions.

The standard thermodynamic diagram used in contemporary meteorology was developed by meteorologists of the United States Air Force and is called the **Skew-T, Log P Diagram**. The diagram is useful for forecasting convection, clouds, fog, types of winter precipitation, and often the surface high and low temperatures.

Figure 17-1. The Skew-T, Log P Thermodynamic Diagram. The five sets of lines are the logarithm of the isobars (log p), the skewed isotherms (skew-t), the dry adiabats, the moist adiabats (sometimes called the saturation adiabats) and the mixing ratio lines. These are explained on the next page. When air parcels are unsaturated and rising, they rise and cool along the dry adiabat upward; when the parcels are saturated after cooling to their dewpoint temperature, they rise (and continue cooling) along a moist adiabat.

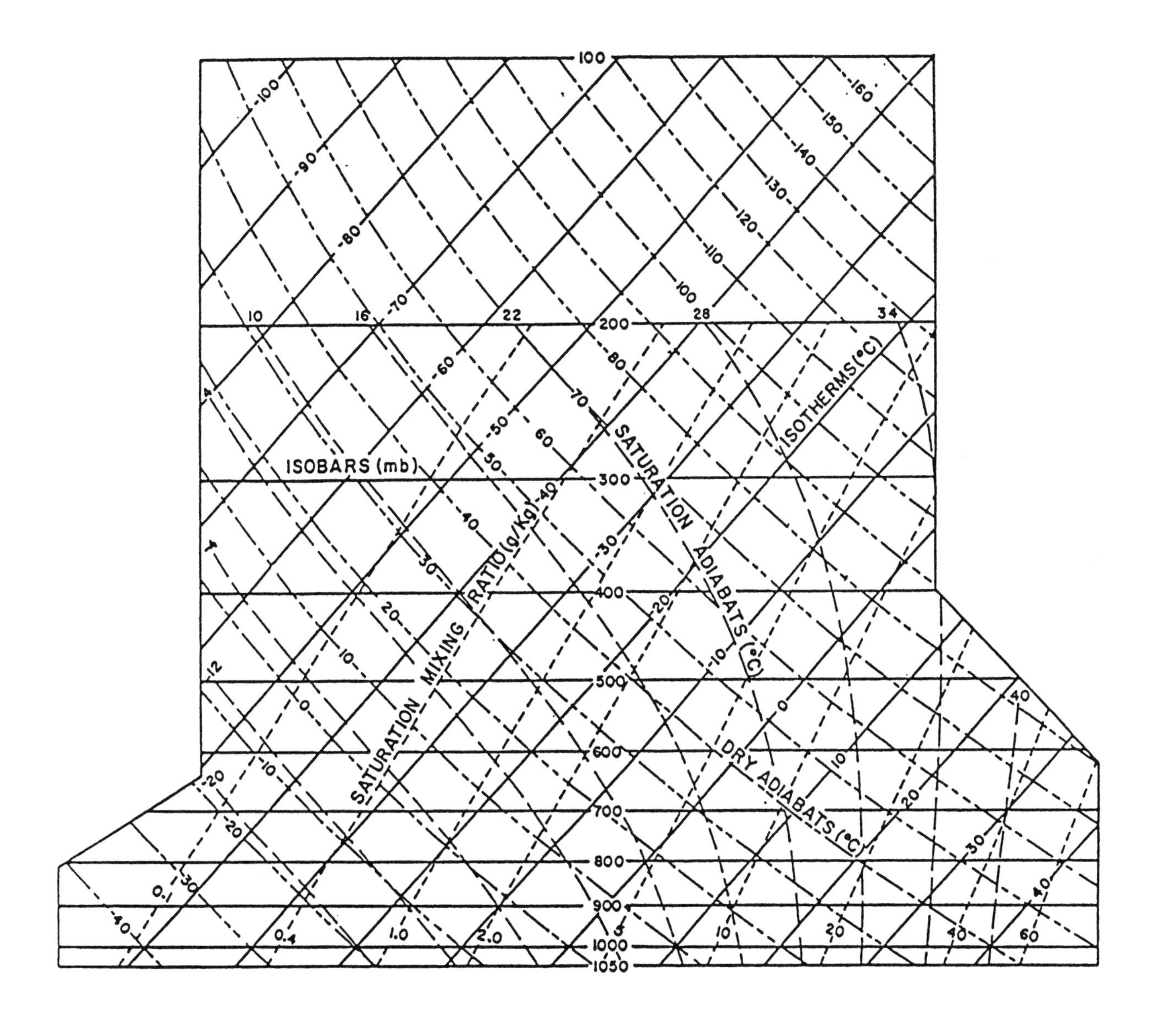

The solid horizontal lines are the pressure levels and the dashed lines sloping rightward are the lines of equal temperature called isotherms. The curved, leftward-sloping lines are dry adiabats, which is the rate of cooling of unsaturated air parcels as they rise, or warming as they descend. These are labelled "dry adiabats". The moist or saturation adiabats, which are labelled "saturation adiabats" do not slope as greatly as do the dry adiabats. As we know, air parcels rise along or parallel to a dry adiabat when unsaturated and along a moist adiabat, labelled "saturation adiabat", when saturated (i.e., when they have risen enough to cool to their dewpoints).

Across the bottom of the chart, the isotherms are labelled in degrees Celsius. On the right end of the thermodynamic diagram (not shown in figure 17-1) are plotted the wind directions and speeds at reportable heights, with the speed in knots. This will be shown in an example of a plotted sounding with wind information, later.

Now let us look at a plotted sounding. We have received data from either a radiosonde or satellite sounding. The temperatures and dewpoints at the various heights or pressure levels are then plotted on this diagram and the data points for the temperature profile in the vertical are connected, and the data lines for the dewpoint temperature profile in the vertical are connected.

The mixing ratio lines give the parts per thousand, actually in grams per kilogram, of water vapor that the air is holding for that dewpoint at that pressure. If the air held all the moisture it could for that temperature and pressure, then the dewpoint and temperature would be the same. This is how we compute relative humidity. Take the surface mixing ratio value that the dewpoint would give, and divide it by the mixing ratio value for the temperature, and multiply the result by 100% to express the relative humidity in percent of saturation for that temperature.

Air conserves most of its dewpoint (moisture content) as it ascends, if indeed some **forcing mechanism** is causing the air to rise. That is to say, air parcels keep or conserve their dewpoint or moisture content, for the most part, as they ascend, with only a slight dewpoint lapse rate with height. The atmospheric pressure is the sum of the partial pressure of each of its constituents. Water vapor is one of these atmospheric constituents, and its pressure is affected as it rises into lower environmental pressures. Thus, we can say that air conserves its dewpoint as these air bubbles rise, but in reality there is a slight lowering of the dewpoint during such rising.

If the rising air is not saturated, it cools along the dry adiabat (at the dry adiabatic lapse rate of about 9.8 Celsius degrees per kilometer). When saturated, it then cools at a lower rate since the rising air parcels are then releasing the **heat of condensation** as clouds and subsequent precipitation are formed. These parcels then cool at about 6 Celsius degrees per kilometer, but that drops to about 3 Celsius degrees per kilometer in the higher troposphere, and after virtually all of the moisture has been condensed ("squeezed") out, the air resumes cooling at the dry adiabatic lapse rate if it keeps rising.

The dry and moist adiabatic lapse rates are the rates at which unsaturated, and saturated, respectively, parcels of air cool as they ascend. The term, **adiabatic**, means that during a physical process, such as air parcels rising on their own because they are positively buoyant, no heat is added to or taken away from the air parcel. In reality, very meager heat is generated through friction of the air molecules and larger parcels and by entrainment of some environmental air, but this is considered meteorologically insignificant. As long as the rising bubbles of air, about a cubic meter in volume, are warmer than the environment, they keep rising.

Figure 17-2. Below is a plotted sounding. The solid line sounding is the temperature curve gotten from data from an ascending weather balloon and/or from a sounder on a weather satellite. The dashed line sounding is the dewpoint curve. Across the bottom of each sounding is the temperature scale, in degrees Celsius. The logarithm of pressure, in millibars (hectoPascals) is the vertical ordinate.

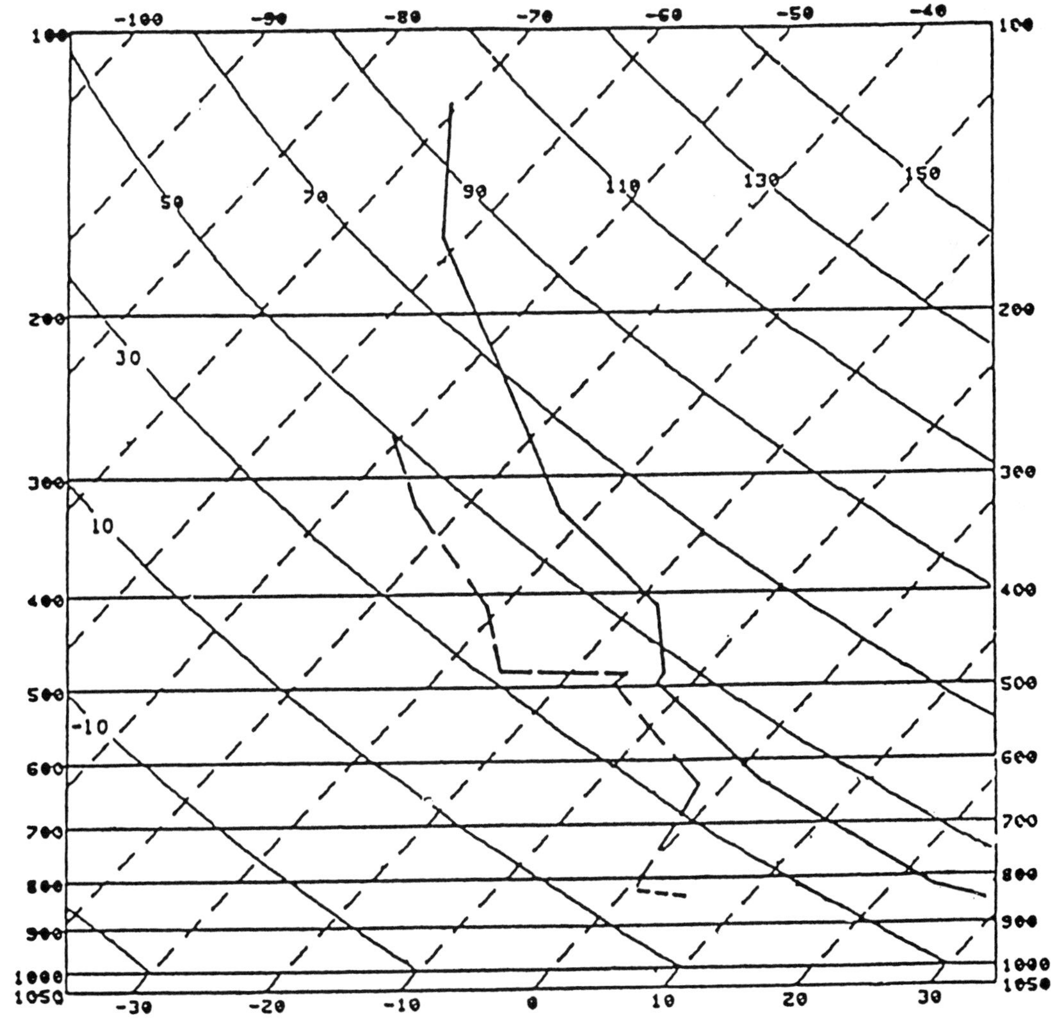

Notice how the temperature lines are not vertical but are skewed to the right. This makes it easier to read, and meteorologists can also use such a chart to compute energy related to rising and sinking air parcels.

As stated earlier, the skew-T diagram is used most often in convective forecasting. Refer to the following chapter, chapter 18, for information on the most-used convective indices, the Lifted Index, Showalter Lifted Index and the K-Index, which can be computed from the skew-T diagram. Moreover, the precipitable water overhead comes from the sounding plotted on this diagram. This is the amount of water vapor in a column from the surface to about 500 mllibars; this is how much would precipitate out if it all fell on you. However, we know that in thunderstorms, the converging winds bring in more moisture from the surrounding area, so that the actual rainfall potential may be from three to five to sometimes as much as seven times the precipitable water value.

The strength of updrafts can be qualitatively inferred by analyzing the convective available potential energy (called "CAPE") from a sounding on the skew-T. Since what separates a routine or ordinary thunderstorm from a severe thunderstorm is the strength of the updrafts, and since tornadoes often occur in storms that are producing large hail, and since large hail occurs only with powerful updrafts, it is therefore useful to be able to estimate the strength of updrafts from a sounding or predictive sounding. (The current sounding does not necessarily show the approximate conditions several hours later; thus, forecasters will estimate low-level temperatures, dewpoints and winds, and use this data to modify the original sounding. Then, a timely analysis and prognosis are done.)

One of the prognoses done is to determine how much parcels will rise above the equilibrium level. The equilibrium level is the level at which the rising air parcels eventually become the same temperature as the environment and stop accelerating. They keep rising, but at a decelerating pace until they stop.

Now let us look at a plotted sounding (next page), showing the temperature and dewpoint plots, as well as the winds (directions plus speeds in knots) on the right-hand side of the chart. Our analysis, which is a modified or prognostic sounding, was done on a computer workstation using a skew-T program.

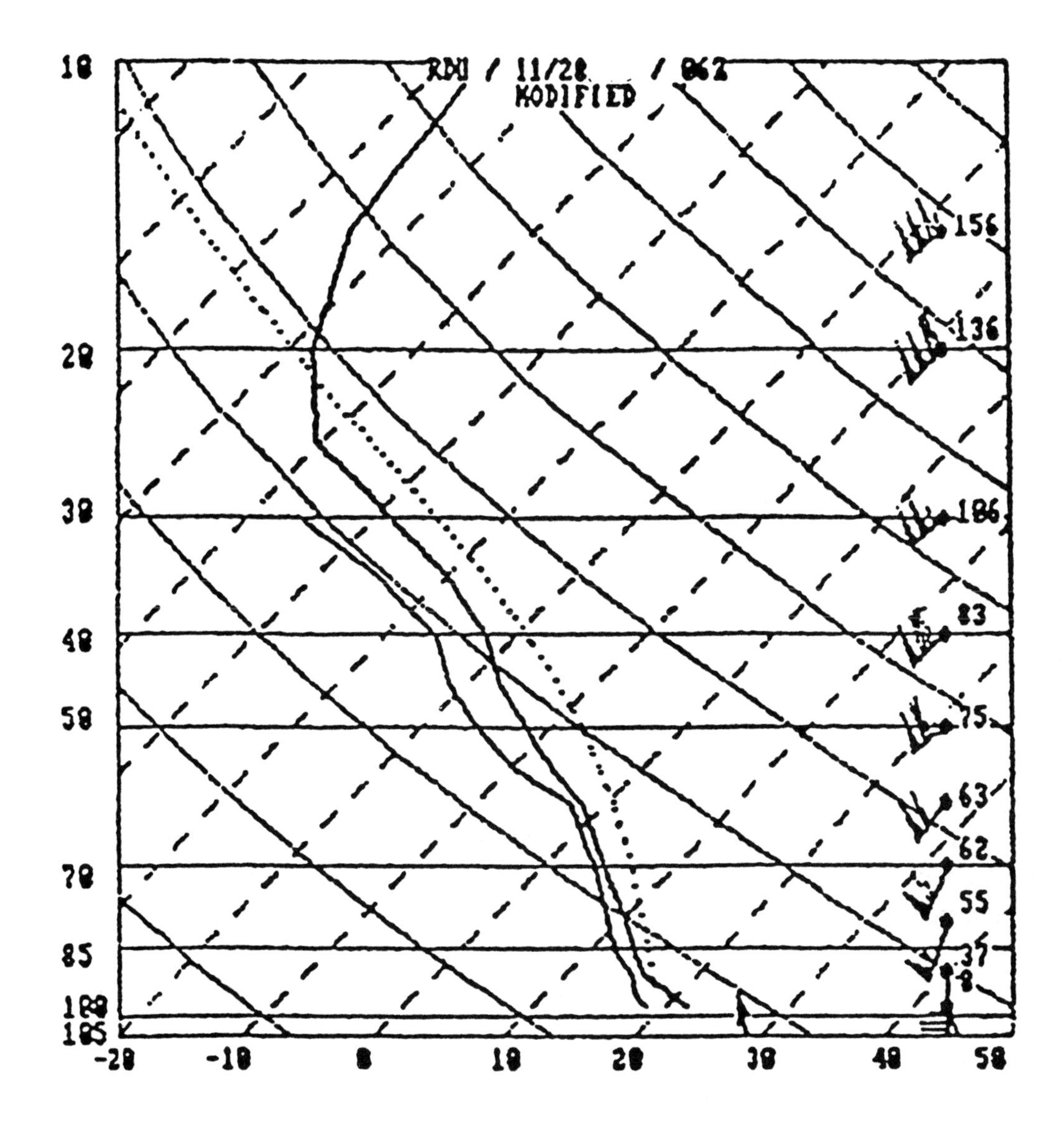

Figure 17-3. A sounding showing the path of the rising air parcels and the equilibrium level. This skew-T, log P diagram shows the following. At left is the pressure level, with the last zero left off, in millibars (mb). Thus, 85 means 850 mb, 70 means 700 mb, etc. This is the height to which air must rise to reach that pressure level in the atmosphere. As an approximation, for a rough average, 850 mb is about 5000 feet up, 700 mb is about 10,000 feet elevation, 500 mb is at about 18,000 feet, 300 mb is about 29,000 feet, 200 mb is about 39,000 feet and 100 mb is at about 53,000 feet...about 10 miles...up. (source: NWS)

Look at the two solid black lines extending from the surface up. The right line extends to 100 mb, while the line to its left stops at 300 mb. The right line is the temperature sounding. The left curve is the dewpoint sounding. Thus, at 300 mb, which is about 29,000 feet up, the temperature is -40°C and the dewpoint is about -46°C.

The CAPE (convective available potential energy) is determined from the positively buoyant area. (continued)

The dotted curve from the surface on up is the path of rising air parcels. Since the parcels have cooled to their dewpoints at a low elevation, they rise and cool at the moist adiabatic lapse rate, and above about 300 mb, when most of the moisture has been expended, they then cool at close to the dry adiabatic lapse rate from that point and farther up.

Notice that as long as the parcels are warmer than the environment, they keep rising. That is, if the dotted line is to the right of the temperature sounding, the parcels are positively buoyant, i.e., they are warmer than the environment, and they keep rising. They actually keep accelerating, going faster and faster, until they hit the equilibrium level, which is the level at which the parcels of air are at the same temperature as the environment, i.e., at the same temperature as that shown by the temperature sounding line. This happens where the dotted line crosses the temperature sounding line, which is at about 220 millibars in our example. Thus, the equilibrium level at this time on this day at this location, which is Raleigh-Durham, North Carolina, is at the 220 mb pressure level.

Since the parcels have been accelerating, that is, increasing their vertical speed, as they ascent up to the equilibrium level, then the equilibrium level also is where the greatest upward vertical velocity of the parcels can be expected. Then, as parcels rise above the equilibrium level, notice that the dotted line shows them to be colder than the environmental temperature. Now, they are decelerating, slowing down. By the time they reach to about 100 mb, they will have stopped rising.

How high will the parcels go? Meteorologists shade in the area from the equilibrium level (EL) on down, between the temperature sounding and the rising parcel path, and call this the **positively buoyant energy area**. We then shade in the area above the equilibrium level from the temperature sounding to the rising parcel path line until that shaded area equals the shaded areas of the positively buoyant energy area. The area above the EL is the **negatively buoyant energy area**. Thus, when the negatively buoyant energy area equals the positively buoyant energy area, the parcels stop their ascent, and this is about how high the thunderstorm tops will grow to. In time, these cloud tops tend to sag back down towards the equilibrium level, especially as the updrafts weaken as the thunderstorm passes its mature stage and enters the dissipation stage.

CAPE is the area between the environmental temperature curve and the dotted curve which indicates the path of the parcels. It is the positive buoyant area. The more energy available, the greater the likelihood of having strong vs. weak thunderstorms.

An in-depth explanation of all the uses of the skew-T diagram would require a book itself! Fortunately, there is an excellent operational text on the subject, entitled, "Use of the Skew-T, Log P Diagram in Analysis and Forecasting", published by the United States Air Force.

Figure 17-4. Use this figure of an analyzed sounding with the descriptions of convective operations on the next page.

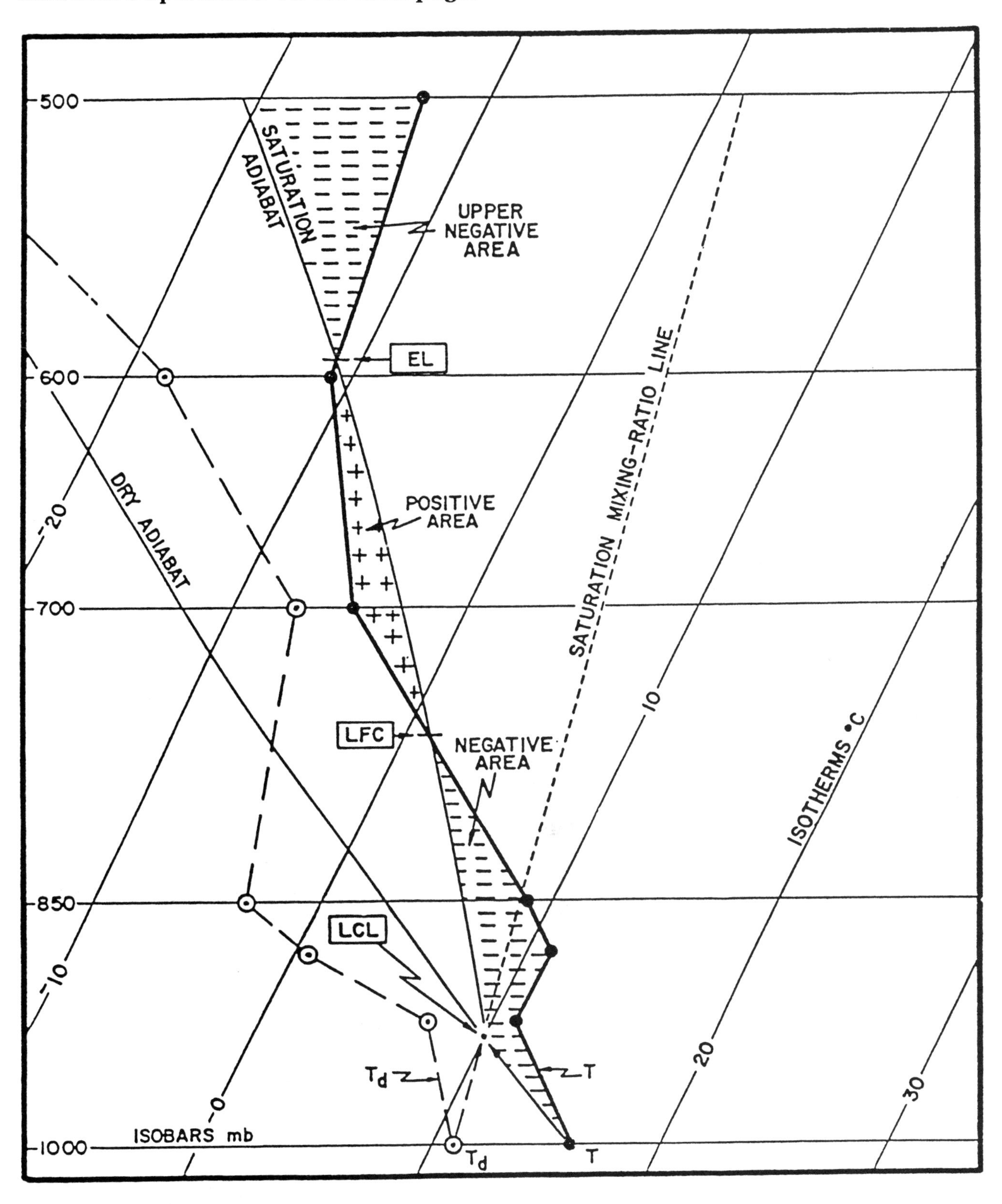

LCL -- LIFTED CONDENSATION LEVEL: Extend a line parallel to the dry adiabats upward from the surface temperature. Extend another line parallel to the saturation mixing ratio lines upward from the surface dew point. Where they cross is the LCL. This indicates where a parcel of air, lifted from the surface dry adiabatically, will become saturated. Often, if surface values are not reflective of the temperature profile or moisture content of the entire surface layer (Within about 1 km) then a mean temperature and mean mixing ratio value may be chosen for determining the LCL.

LFC -- LEVEL of FREE CONVECTION: From the LCL extend a line upward parallel to the saturation adiabats until it intersects the plotted temperature curve. That point is the LFC. Under very stable atmospheric conditions, it is possible to have no LFC (the plotted temperature curve would be warmer than, or to the right of, the saturation adiabat of the LFC). The Level of Free Convection is the point where a surface-based parcel, lifted to saturation and beyond, is able to rise freely because of its positive buoyancy.

EL -- EQUILIBRIUM LEVEL: From the LFC, draw a line upward parallel to the saturation adiabats (this line MUST BE to the right of, or warmer than, the plotted temperature curve). Obviously, if there is no LFC for the sounding, there can be no EL. The point where this line crosses the temperature curve and stays to the left of the temperature for the remainder of the sounding, is the Equilibrium Level. The line may make one or more temporary crosses to the left of the temperature due to mid-level inversions, or layers of relatively more stable air, but the last point where the line crosses permanently to the left is the EL. The Equilibrium Level divides the sounding into areas of (predominantly) positive buoyancy below, and negative buoyancy above. At that level, a rising parcel of saturated air becomes colder than the environmental temperature and begins to decelerate. The parcel will theoretically come to a halt at the MPL.

MPL -- MAXIMUM PARCEL LEVEL: Continue the line from the LCL, LFC, and EL upward along the saturated adiabats (if above 200MB, follow the dry adiabats) until the amount of negative buoyancy, in area, is equal to the amount of positive buoyancy, starting at the LFC. If a parcel, lifted from surface, saturated, and brought to the LFC by lifting, rises moist adiabatically, remains saturated through the entire ascent, is not subject to entrainment or slowed by the weight of condensed water above it (left by preceding parcels), it will come to a gradual halt at the MPL.

TROP -- TROPOPAUSE: Find the lowest point in the sounding where the temperature lapse rate decreases to 2°C/km and remains so for a distance of 2 km. The average lapse rate within this 2 km stratum cannot exceed 2°C/km. The Tropopause is the boundary layer between the troposphere below and the stratosphere above.

Chapter 18. INSTABILITY INDICES

Over the years, weather forecasters have invented indices based on sound physical reasoning, that are helpful in convective forecasting. Three of these experimental indices have proven to be the most useful and are now standards in convective analyses: the lifted index, also known as the LI, the Showaltar Lifted Index, which is named after its creator, and the K-Index. The LI and Showaltar LI are useful for thunderstorm and severe thunderstorm probability forecasting, and the K-Index is useful for predicting thunderstorms with heavy rain and flash flooding potential.

The lifted index (LI)follows the parcels as they rise, and at an arbitrary level, 500 mb, looks at the temperature of the sounding and the temperature of the parcel. The 500 millibar level is used since it is about 18,000 feet high and in the middle, more-or-less, or within the most "meaty" area of most thunderstorms.

The Lifted Index (LI) = the sounding's (environmental) temperature MINUS
the parcel's temperature

If the parcel is warmer than the environment it is rising through, then the LI is a negative number, which refers to an unstable environment.

If the parcel were to be forced up to 500 mb but is colder than the environment, then we have a positive number, which refers to a stable environment. In a stable environment, the parcel will be negatively buoyant; therefore, it would sink, not rise.

A stable sounding can become unstable by making ITS lapse rate increase. This can be done by warming the environment in the low-levels of the troposphere and/or cooling it aloft, also by adding moisture in the lower levels and/or drying it aloft.

Thus, a +2 LI, which is minimally stable, could change to, e.g., a -2 LI in say 6 to 12 hours. An example is a sunny day with a morning +2 LI becoming a mid-afternoon -2 because solar heating has warmed the ground which in turn warms the lower layer of the troposphere. This is assuming no significant low-level warm or cold air advection which would also contribute in a destabilizing or stabilizing way, respectively, to the environmental sounding.

Thus, by warming in low-levels by heating of the day, for example, or by warm air advection, and/or by cooling aloft, the slope of the temperature sounding steepens. The temperature drops off faster with height. This makes the sounding less stable. Keep in mind that the rising parcels of air can cool at only the dry adiabatic lapse rate, and when saturated, at the moist adiabatic lapse rate. These rates do not change. However, the environmental lapse rate does change. This is how destabilization and stabilization occur.

Reviewing, then, what the lifted index process is: we raise air from or near the surface to 500 mb and compare the rising parcel's temperature there with the 500 mb temperature. If the parcel is warmer than the 500 mb environmental temperature, then we have instability and a negative lifted index value; if the parcel is colder than the 500 mb environmental temperature, then we have stability and a positive lifted index value. The lifted index is the 500 mb temperature minus the temperature the parcel would have if lifted to 500 mb. The air parcel cools as it rises, at the dry adiabatic lapse rate if the parcel is initially unsaturated, and at the moist adiabatic lapse rate once the parcel cools to its dewpoint temperature, becoming saturated, or if the parcel is initially saturated when it starts rising.

The rising or lifting is caused by some lifting mechanism, as discussed in earlier chapters of this book. By way of review, some examples of lifting mechanisms are:

- solar heating during the daytime if the sky is clear of at least most non-thin clouds;
- an advancing cold front which lifts the warmer air ahead of it;
- an OUTFLOW BOUNDARY from a thunderstorm or remnant of a thunderstorm or thunderstorm complex;
- other advancing boundaries such as a sea-breeze front;
- air flowing up a rising terrain, such as UPSLOPE or, for up mountains, OROGRAPHIC LIFTING;
- intersection of boundaries...e.g, if two thunderstorms approach each other, their intersecting outflows result in major mass (of air) convergence and new lift which forces new convective development.

If there are no big temperature inversion layers below 500 mb, then the following lifted index values are used as a guide:

LIFTED INDEX	CONDITION
+6 or higher	quite stable
+3 to +5	stable
0 to +2	weakly stable
less than 0	unstable
-6 or less	quite unstable; greatest tornado potential

The values for the Showaltar LI are the same as for the LI, but to determine the Showaltar LI, you start at 850 mb, raise the parcels to their lifted condensation level, and then continue to 500 mb as with the LI. The theory for starting at 850 mb rather than at the surface or rather than taking an average temperatures and dewpoint for the first 50 mb off the surface and starting with that, is that you are well above the planetary boundary layer so that local affects, especially a temperature inversion, will not throw off the LI. Consider this: an inversion, which may exist say in a morning sounding after a clear, nearly-calm night, would most often be broken within a few hours after sunrise, but if the index were computed using the readings from in the inversion, the LI would not be representive of the convective potential of the local troposphere.

LIs can be misleading in high mountainous regions, where the surface pressure is much lower than 850 millibars. In such a case, going above the 500 mb level, say to the 400 mb level, may yield values that are more representative of the local; troposphere's convective potential.

The K-Index (KI) (the "K" does not stand for anything) is an invention used to predict thunderstorm potential and heavy rain/flash flood potential.

The K-Index = (850 TEMP - 500 TEMP) + 850 DEWPT - 700 DEPRESSION

TEMP is the temperature, dewpt is the dewpoint, and depression is the difference between the temperature and the dewpoint; 850 is 850 mb, 700 is 700 mb and 500 is 500 mb; all of these temperatures are in degrees Celsius.

The first term of the KI formula tells how steep the environmental lapse rate is between 850 and 500 millibars. The greater the temperature drop through that layer, the lower the stability. A good value for an unstable environment is 30 or more.

The second term is the 850 mb dewpoint temperature. If we are looking for heavy rain, we want as high an 850 mb dewpoint as possible. A warm season value of 10 or more would be high.

For flash flooding, we need slow-moving organized convection with deep moisture. The third term is the 700 mb temperature/dewpoint depression or spread, i.e., the difference between the temp. and the dewpoint at 700 mb. For heavy rain, we need copious amounts of moisture..water vapor..through a deep layer. Thus, the dewpoint should be high from the planetary boundary layer up through about 500 mb. Therefore, at 700 mb, the closer the temperature and dewpoint are to each other, the better for heavy rain potential when the first two terms of the K-Index are also high. Zero, then, is the ideal value for the third term.

Consider our example of 30 for term 1 and 10 for term 2. If zero is the value for term 3, then our KI is 30+10-0=40. If the air were very dry at 700 mb, e.g., if the depression there were 20, then the KI would be 30+10-20=20. The higher the KI, the greater the threat of thunderstorms.

Here is the chart of KI values are their relevancy for thunderstorms and their heavy rain potential.

Values used in operational forecasting are:

WHEN THERE IS ADEQUATE LIFTING OF THE PARCELS:

K-Index	Relationship to Convection
<28	thunderstorms not likely
28 to 32	chance of thunderstorms
33 to 35	good chance/likely to have thunderstorms
36 or above	expect thunderstorms
38 or above	heavy rain/flash flood potential

A KI of 40 or above is considered "extremely juicy". It would be accompanied by rather high precipitable water (PW) values (see chapter 17). PWs well in excess of one inch would likely accompany a KI of 40 or more. PWs can reach 2" or more with a very moist dewpoint sounding.

CAVEAT: Sometimes a high KI can be misleading. Since we are looking at values at only three levels, 850, 700 and 500 mb, sometimes there can be dry slots between these levels. If we were to take these into account, there would not be deep, continuous high moisture content.

To check to make sure that the dewpoints do not dry out between these levels, it is necessary to look at the weather balloon sounding of temperature and dewpoint in the vertical, or to look at the radiosonde weather report itself to read these values and assure that the atmosphere is indeed moist. With dry slots aloft there may still be thunderstorms, but they would likely not be heavy rain producers and flash flood threats. Moreover, dry intrusions aloft around 700 mb actually contribute to making the thunderstorms more severe, since the temp. and dewpoint lapse rate may be greater which allows the rising parcels to accelerate faster and higher, leading to more violent updrafts. What separates a severe thunderstorm from a routine one is the strength of the updrafts.

Chapter 19. CONVECTIVE AVAILABLE POTENTIAL ENERGY (CAPE)

Figure 19-1. The sounding at the right shows the positive buoyancy area whose energy potential for the rising parcels can be computed (the CAPE, or Convective Available Potential Energy) and used as a measure of how unstable the local environment is. In general, when the conditions for thunderstorm development exist, then the higher the CAPE value, the more likely are the storms to be severe. Large hail is usually associated with high values of CAPE.

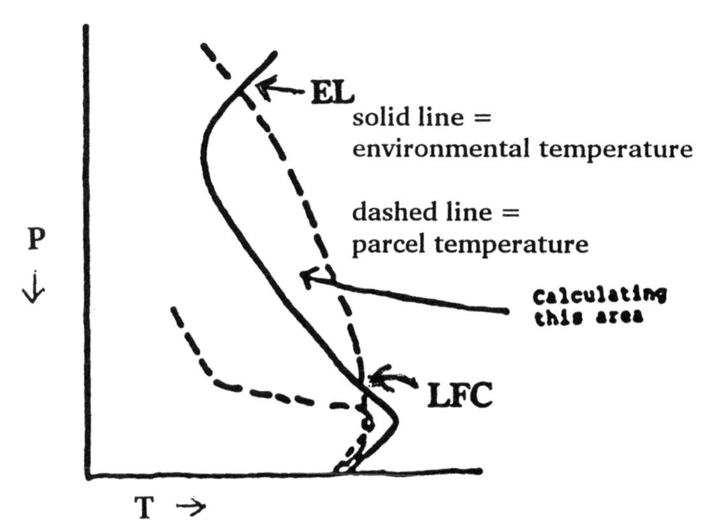

Thus, CAPE represents the amount of available potential energy in the area that is buoyant, which is the area between the level of free convection (LFC) and the equilibrium level (EL). When parcels are raised to the LFC, then they no longer need additional lift supplied to them because they are then warmer than the environment (buoyant) and will accelerate upwards on their own.

By increasing the moisture in parcels of air, that is, replacing some of the dry air with water vapor, the air becomes more buoyant; and by warming the parcels, the air becomes more buoyant. When air is warmed and moistened, its stability decreases, that is to say, the environment becomes more unstable (the parcels ultimately may become buoyant and can rise readily through the region between the LFC and the EL).

Recall from the gas laws that water vapor weighs less than dry air, so replacing some of the dry air with water vapor lowers the mean molecular weight, making the parcels lighter and therefore more buoyant.

The equation of state shows that if P (pressure), V (volume) and T (temperature) remain constant along with R (the gas constant) which is constant, then m (the number of molecules in the volume) must also remain constant. This means that if a molecule of water vapor enters the volume, then a molecule of dry air must exit the volume for P, V and T to remain constant. The mean molecular weight of water is about 18 g/mol, while the mean molecular weight of dry air is about 29 g/mol. Thus, an air parcel that contains water vapor weighs less (is more buoyant) than a similar parcel of air that contains no or fewer water vapor but has the same pressure, volume and temperature.

In a sense, then, moisture and instability are closely related. This is where CAPE comes in. A single value can describe the buoyancy that a parcel has when both temperature and moisture of the rising parcel and of the environment are both taken into account. This number is the CAPE.

Usually, the "virtual temperature" is used in this thermodynamic equation, rather than the actual temperatures of the parcels and environment. The virtual temperature is the temperature that dry air would have if its density were equal to that of a given sample of moist air at the same pressure.

$$Virtual\ Temperature = T_v = T\,\frac{1 + \frac{w}{\epsilon}}{1 + w}$$

w = mixing ratio (kg H_2O/kg dry air)

T = air temperature (K)

$$\epsilon = 0.622 = \frac{molecular\ wt\ of\ water}{molecular\ wt\ of\ dry\ air}$$

The actual temperatures, however, could be used, since the difference between the temperature and virtual temperature is that the virtual temperature may be from about 1 to about 7 degrees Kelvin warmer than the air temperature.

A formula for the CAPE is (B+ refers to positive buoyancy):

$$CAPE = B+ = R_d \int_{P_{lfc}}^{P_{el}} (T_{v_p} - T_{v_e})\, d\ln(P) = \frac{m^2}{sec^2}$$

R_d = dry air gas constant

P_{lfc} = pressure at the level of free convection

P_{el} = pressure at the equilbrium level

T_{vp} = virtual temperature of rising parcel (K)

T_{ve} = virtual temperature of the environment (K)

Thus, from the formula, we see that we are summing up or integrating from the LFC to the EL the positive energy area of a lifted parcel on a sounding. This is the CAPE.

<u>How to use values of CAPE</u>:

Weather researchers have found that the higher the CAPE values, the more likely the threat of thunderstorms, and that very high CAPE values increase the risk of severe thunderstorms. The table below summarizes the relationship of CAPE values and thunderstorms:

CAPE value: (in Joules of energy per kilogram of mass)	Forecast implication:
under 1000	weakly buoyant environment
1500 to 2500	moderately unstable; thunderstorms likely if no strong cap aloft
greater than 3500	extremely unstable; severe thunderstorms possible

Keep in mind that CAPE alone may not tell us everything about influences on the vertical motion. Powerful updrafts can occur even in weakly buoyant environments when strong enough vertical wind shear exists from the surface through the mid-troposphere. Such wind shear with very high CAPE values is especially dangerous. What happens is that the updrafts are actually increased due to the interaction of the updrafts and the vertically-sheared environment: the air is forced to spin up more rapidly.

Some researchers have tied CAPE to estimating the maximum possible upward vertical motion, which would occur at the EL since air is accelerating from the LFC to the EL, and after the parcels rise above the El, they are cooler than the environment and begin decelerating until they stop their ascent (which is the MPL or maximum parcel level). You can estimate the maximum speed (in meters per second) of the updraft for a parcel rising adiabatically by converting CAPE to kinetic energy using the following expression:

$$W_{\max} = \sqrt{2 \times CAPE} = \frac{m}{\sec}$$

Actual values of the updraft maximum speed (W_{max}) are less than the computed value due to water loading in the updraft and the effects of mixing of air and entrainment of some drier air from just outside the storm. However, in well-organized storms such as supercells, the actual updraft maximums are close to this computation since entrainment in such storms is minimized, the storm-relative flow tends to carry much of the precipitation away from the updraft summits, which lowers the water-loading affect that slows the updraft somewhat, and other dynamical forces such as significant wind shear with height enhances the speed of the updraft. An example of a very high CAPE value and its computed maximum updraft speed: a CAPE of 3000 Joules/kg can produce a maximum upward vertical velocity, $W_{max} = (2 \times 3000)^{1/2} = 77.5$ meters/second which is about 155 knots or nearly 180 miles per hour!

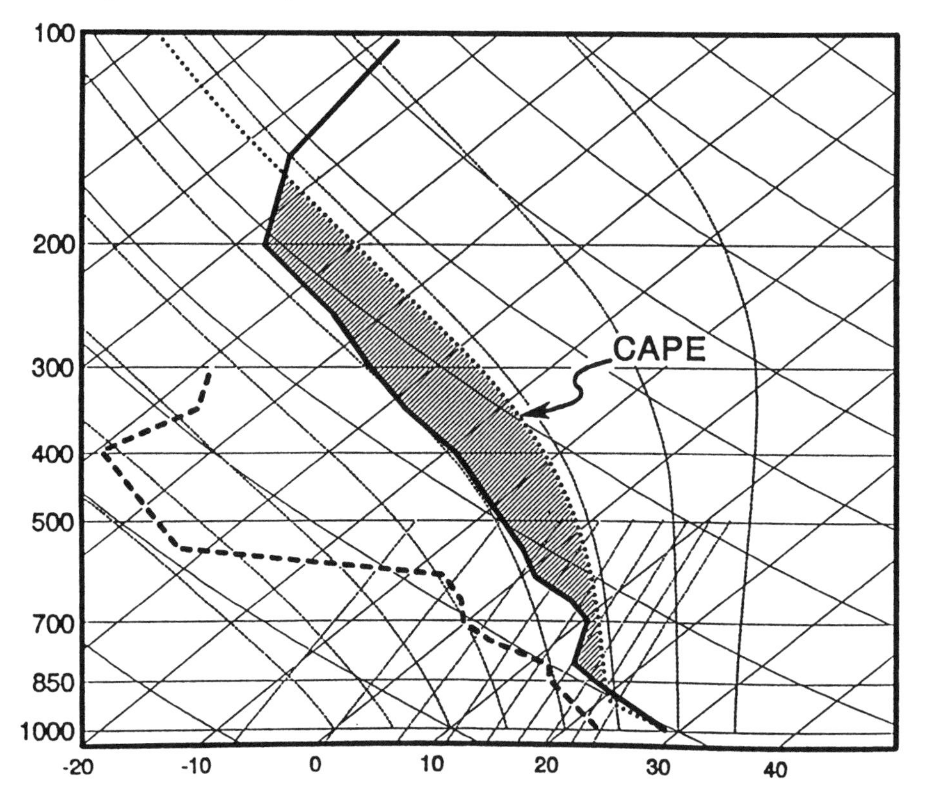

Lifted Index	-6
CAPE	3020 Jkg^{-1}
Equilibrium Level	43,000 ft
Precipitable Water	1.64 in
LCL	2,417 ft

Figure 19-2. **An analyzed sounding showing a lifted index of -6, which is very unstable with the risk of tornadic thunderstorms, and a very high CAPE of 3020 Joules per kilogram.** With no strong mid-level cap showing on the sounding, and the EL way up at about 160 mb, into the lower stratosphere, this is a "loaded gun" sounding for potential violent thunderstorms to develop, once air is forced up to the LFC, Note that around 700 mb there is a cap inversion, but it is not quite warm enough to intersect the path of the rising parcels to stop them. Moreover, if this were a morning (around daybreak) sounding, the forecasters would have to make a predictive sounding for how the lowest part of the sounding would look by early afternoon so see if the CAPE value would be even higher in warmer low-level air and perhaps an increase in the low-level moisture. Surface moisture flux convergence would bring in additional air (convergence), which would carry in additional water vapor to the local environment. Recall that the four ways to destabilize the local troposphere are: low-level warm air advection, low-level moisture advection, mid-level (meaning mid-tropospheric) cold air advection and mid-level drier air advection. (source: NOAA)

The negative buoyancy of the negatively buoyant area from the surface to the LFC can also be computed, to determine how much lift is required to force the parcels to the LFC. An expression for the negative buoyancy, B-, is:

$$B- = R_d \int_{P_{sfc}}^{P_{lfc}} (T_{v_p} - T_{v_e})\, d\ln(P) = \frac{m^2}{\sec^2}$$

and for the lift required to get the parcels to the LFC where they would be buoyant, is:

$$W_{lift_{req}} = \sqrt{(-2) \times (B-)} = \frac{m}{\sec}$$

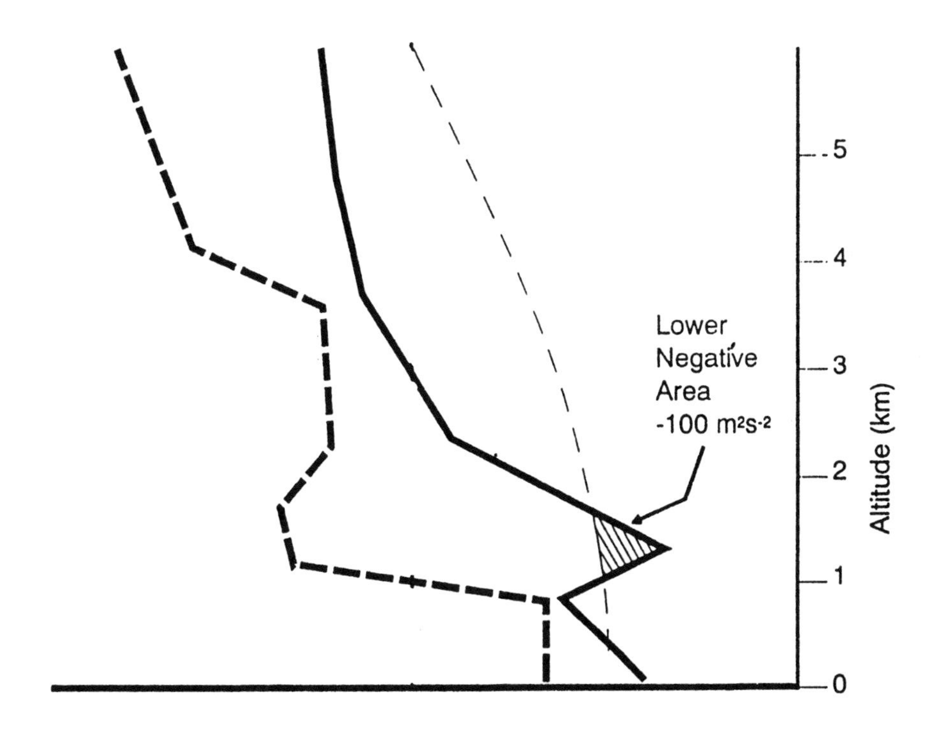

Figure 19-3. The temperature sounding (solid heavy line) and dewpoint sounding (dashed heavy line) and raised parcels path (dashed lighter line) with the hatched-in area being the negatively buoyant area below the LFC. This amount can be converted into the lift required, in meters per second, to force the parcels up to the LFC. (source: NOAA)

Chapter 20. **THE BULK-RICHARDSON NUMBER (BRN)**

The Bulk-Richardson Number or BRN (invented by a researcher named Richardson) is a ratio of the local buoyancy to the local wind shear.

Buoyancy is a critical factor in thunderstorm development: the more buoyant the local atmosphere through the longest vertical fetch, the more likely is thunderstorm development as long as the conditions for convection exist. The right amount of change in wind direction and/or speed (the vertical wind shear), the more likely are any thunderstorms that develop to become severe.

Thus, the BRN is an insightful way to compare the amount of buoyancy to the amount of shear, in an attempt to forecast the type of thunderstorm we might get. This approach has been fruitful, and years of operational use of the BRN have resulted in yet another valuable tool for weather forecasting.

The previous chapter defines mathematically positive buoyancy, which we also defined as CAPE, or Convective Available Potential Energy. This is the numerator in our BRN ratio.

The denominator is wind shear, which we shall call U. The change is wind direction and speed from the surface layer upward can be done in small increments, such as for every 1000-foot layer or every layer of several millibars vertical pressure change. Another approach is to define U as the vertical wind shear given by the density-weighted mean wind speed from the surface up through 6 kilometers, minus the mean wind speed of the lowest ½ km., Our equation for U is therefore:

$$\overline{U} = \overline{U_{6000}} - \overline{U_{500}}$$

Then, the wind shear is defined by ½ U^2, which actually is a measure of the inflow kinetic energy made available to the local storm environment by the vertical wind shear. We can also look at this term as defining in a way the strength of the surface inflow feeding the storm environment and the likelihood of the updraft to become rotational. The buoyancy numerator term of the BRN ratio is a measure of the potential strength of the updraft (and can infer somewhat the potential strength of the downdraft and the boundary layer outflow).

The Bulk-Richardson Number (BRN) expression is:

$$BRN = \frac{B}{(1/2\ U^2)}$$

Notice that the dimensions (units) of both numerator and denominator are in meters squared per second squared (units of energy, or, rigorously, units of specific energy, or energy per unit mass), so that when dividing out the expression we see that the BRN is a dimensionless number.

The full expression is:

BRN = (Convective Available Potential Energy) / (vertical wind shear)

(See chapter 19 for the mathematical expression for CAPE.)

Notice that when the numerator, the positive buoyancy, increases with no change in the denominator, the Bulk-Richardson number increases, and when the denominator, the vertical wind shear, increases with no change in the numerator, the Bulk-Richardson Number decreases. Changes in both numerator and denominator which modify the BRN upward or downward have implications for the kind of thunderstorm activity that may occur when thunderstorm conditions are met.

We have learned in recent years that for a given amount of buoyancy, when we have weak vertical wind shear the local atmosphere tends to produce single cell thunderstorms that are short-lived, but with low to moderate vertical wind shear we get multi-cell thunderstorms, and with moderate to strong vertical wind shear we have supercells or split storms.

The environment must have enough buoyancy to allow convection, and as the vertical wind shear increases, the intensity of the storms that develop tends to increase.

Here are values of the Bulk-Richardson Number and the types of convection that tend to occur with these values:

BRN:	LIKELY CONVECTIVE RESULT:
lower than 10	Severe weather is unlikely since the shear is too strong and convection is blown over
15 to 35	Potential for severe weather, with supercells more likely than multi-cell thunderstorms
higher than 50	multi-cell thunderstorms are likely, since buoyancy is great, leading to more widespread convection which ultimately results in clusters of cells, which merge, and the shear is relatively too low to allow for supercell development

A caveat: a weather forecaster must never use the BRN alone to determine the likely type(s) of convective weather that may follow, but must use all appropriate convective tools when making his/her forecast.

Moreover, the BRN was invented based on studies of thunderstorms in the Plains States of the United States. Some modifications to the usage of the numbers may be necessary if this tool is used elsewhere. Nevertheless, thunderstorms all require the same "ingredients" to occur, so that the basic of convection are true across the planet.

In a study of the most powerful tornadoes, it was found that the BRNs were from 0 to 15, which occur with strong BRN denominators, that is to say, with local environments of strong vertical wind shear. Thus, the BRNs are evidence that a significant change of wind direction with height and of increasing wind speed with height plays a role in the genesis of tornadoes. Inflow to a thunderstorm becomes rotated, which is easy to mentally visualize when you ponder this state of the atmosphere.

Chapter 21. THE HODOGRAPH

The hodograph is a presentation of the wind profile in the vertical. It enables the meteorologist to see the wind direction and speed in the first several thousand feet of the troposphere.

There can be various types of hodograph presentations. We shall look at the hodograph that has been worked into operational weather analysis and forecasting.

All plotted hodograph points are the tips of the wind jvectors from the (x,y) = (0,0) location. The hodograph curve connects the end-points of the wind vectors.

Hodographs show the u (west-to-east) and v (south-to-north) wind components at the different heights from the sounding, with u, v, -u and -v being the aces on the graph. [Example: we have at one level a southwest wind at 5 knots; then the u component is 3 knots and the v component is 4 knots. If we have a northeast wind at say 5 knots, then the u component is -3 knots (the minus sign referring to the direction, i.e., opposite to from the west) and the v component is -4 knots.].

First of all, why create such a graph? Of what practical use is a hodograph? Before we look at some hodographs, consider this: we know that vertical wind shear, that is, the change of wind direction and/or wind speed with height, has a profound effect on the type of thunderstorm that is likely to occur. Suppose we were to study hodographs of environment that give us ordinary thunderstorms, pulse thunderstorms, multi-cell thunderstorms and supercell thunderstorms, for example, and we discover that hodographs for most to all supercells are similar, and that hodographs for multi-cells are similar, and so on for the other thunderstorm types. The we would suspect that we could use this information with pre-storm hodographs to possibly predict the likely thunderstorm type when conditions are ripe for convection. This tool may not work all the time, since other factors than vertical wind shear must of course be analyzed, but the hodograph gives us yet another valuable piece of information in assessing the type of convection to occur. We must also keep in mind that the local environment changes, both before and during convection, and that more than one type of thunderstorm may occur in the same area. Also, one type of thunderstorm, say a multi-cell system, may have one of its cells evolve into a supercell. Thus, a thunderstorm may evolve into another type.

Keeping this in mind, let us look at this convective analysis and forecasting tool known as the hodograph.

Before we look at some hodographs, first consider a plot on polar coordinate paper of the wind vectors:

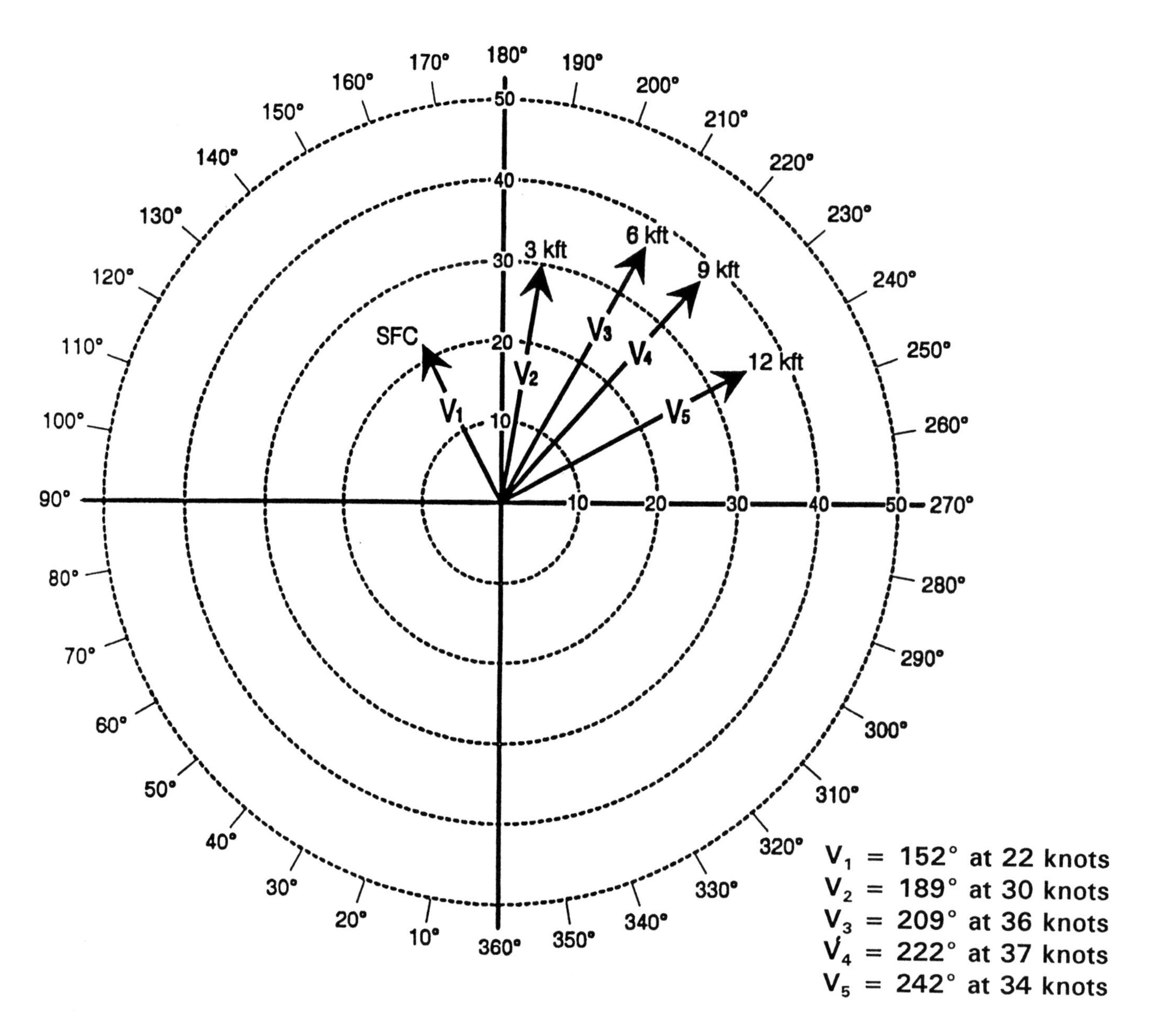

Figure 21-1. A plot of wind vectors on polar coordinate paper. Vector V_1 shows the surface wind, from 152° (from the southeast) at 22 knots. Notice how the rings around the center are in increments of 10 knots in this example, and how the degrees of the compass are reversed from how we normally plot them so that the graph visually shows from which the direction the wind at each level is coming. Vector V2 shows the wind at the 1000-foot level above the ground, V3 is the wind vector for 2000 feet, V4 for 3000 feet and V5 for 4000 feet. This plot is not a hodograph, just a plot of the wind vectors, that is, of both direction and speed, from surface up through 4000 feet for every 1000 feet elevation. This data is gotten by a sounding form a weather balloon or by a wind profiler, which is a Doppler radar aimed straight up to measure the wind. (source: NOAA)

Now let us take this same data and turn it into a hodograph (next page).

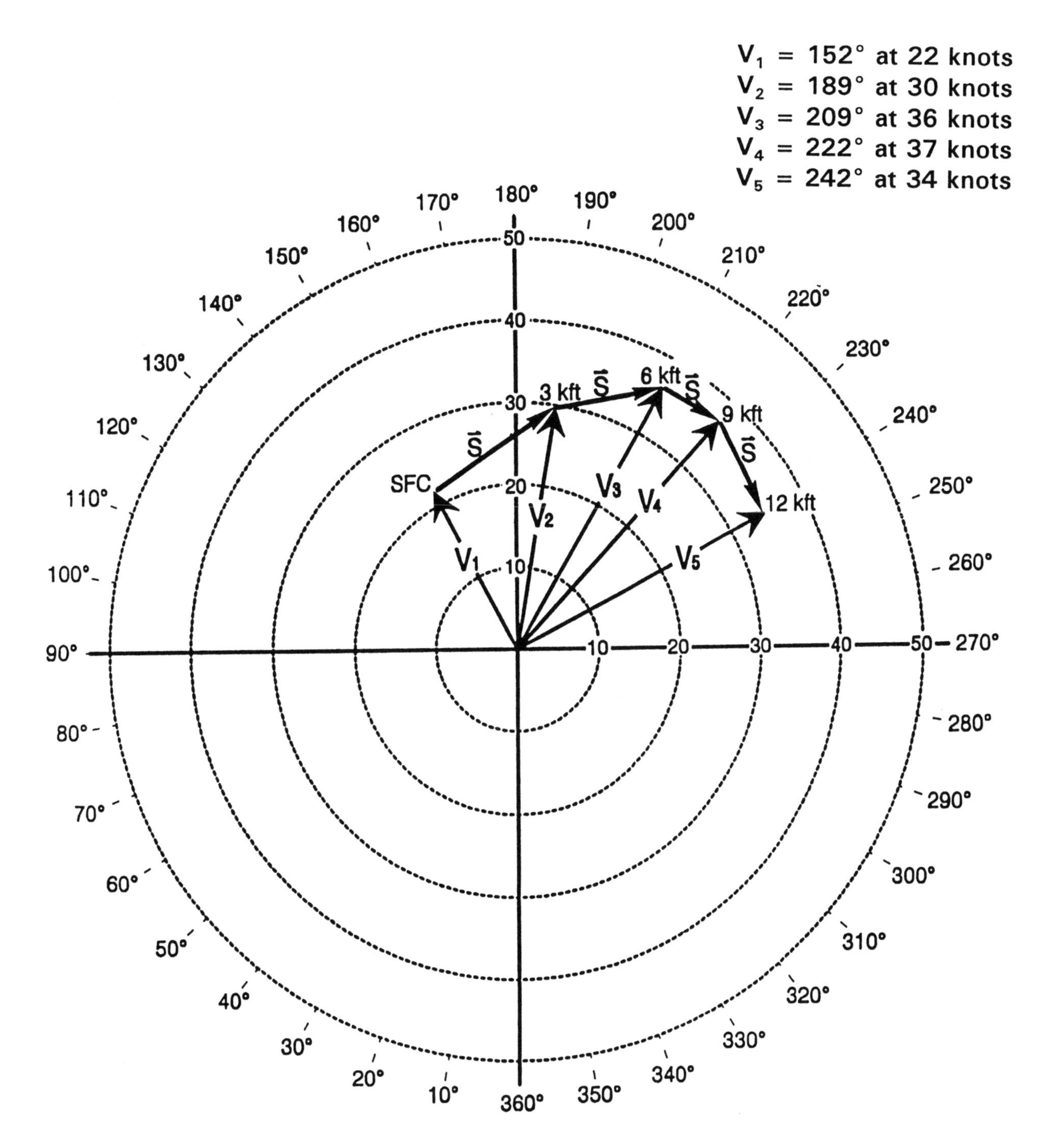

Figure 21-2. An example of a hodograph. Notice that we have connected the end-points or tips of the wind vectors. This curved line is the hodograph plot of the wind profile from the surface up to 4000 feet, in this case. We can continue plotting for every thousand feet, or for other increments, up to 12,000 feet or higher. Again, the fist vector, V_1, is the surface wind, V_2 is the 1000-foot wind, etc., as in the figure on the previous page. (source: NOAA)

Severe weather researchers have found that certain hodograph profiles often predict specific types of thunderstorms when the conditions also exist for convection. The next three figures are from work done on Northern Hemisphere thunderstorms by two researchers, named Chisholm and Renick.

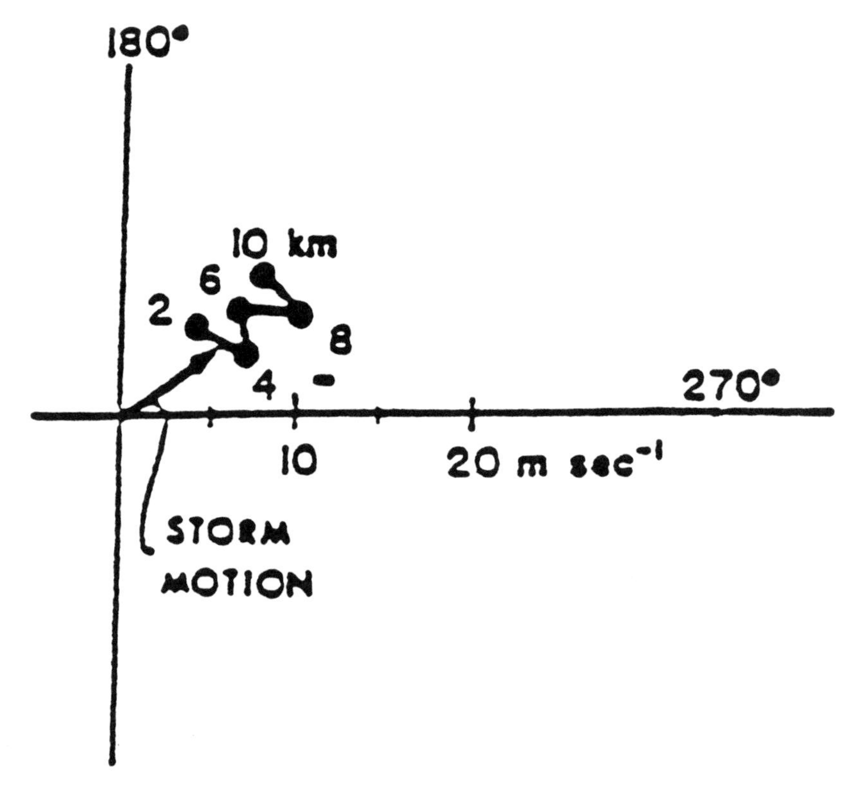

Figure 21-3. A typical hodograph for a Northern Hemisphere air mass or ordinary thunderstorm. In this and the next two examples, the vector end points, which are plotted without the vectors, are labeled in kilometers rather than miles and the speeds are meters per second rather than knots. To convert kilometers to miles, multiply the km value by 0.61, and to convert the meters pr second to knots, double it. Note that for ordinary air mass single-cell thunderstorms, the directional shear is slight and the speed shear is also minor, with light winds. Even at a height of 10 km, which is about 33,000 feet, the u and v wind components are each about 8 or 9 m/sec. (The unlabeled y-axis is also in meters per second; recall the wind speed rings on the polar graph paper, which are not shown here.) The resultant wind, using the Pythagorean theorem, would be the square root of the sum of the squares of the u and v components of the wind, so that in this case the actual wind at 33,000 feet is $(8^2 + 9^2)^{1/2}$ or about 12 meters per second which is about 24 knots or about 28 miles per hour, which is relatively light for so high up. Thus, we have learned that usually the actual wind speed is relatively light, even forso high, in ordinary air mass thunderstorm environments. Little directional and speed shear of the wind is the typical vertical wind profile through the troposphere for air mass, nonsevere thunderstorms. (In the typical case shown above, the storm motion vector is also shown.)

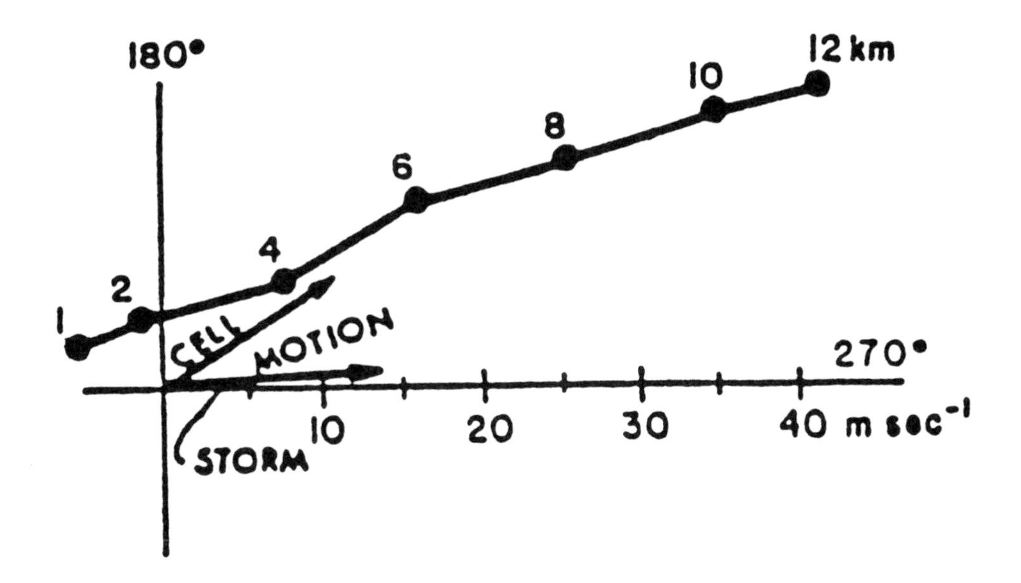

Figure 21-4. A typical hodograph of a Northern Hemisphere multi-cell thunderstorms environment. Notice how the wind direction veers (turns clockwise, left-to-right) in the first 13,000 or so feet (first 4 kilometers on this plot). The wind speeds gradually increase with height. Converting from m/sec, plotted here, to mph, we find that at about 36,000 feet (12 km), u = about 85 mph and v = about 40 mph, so that the resultant wind is near 95 mph.

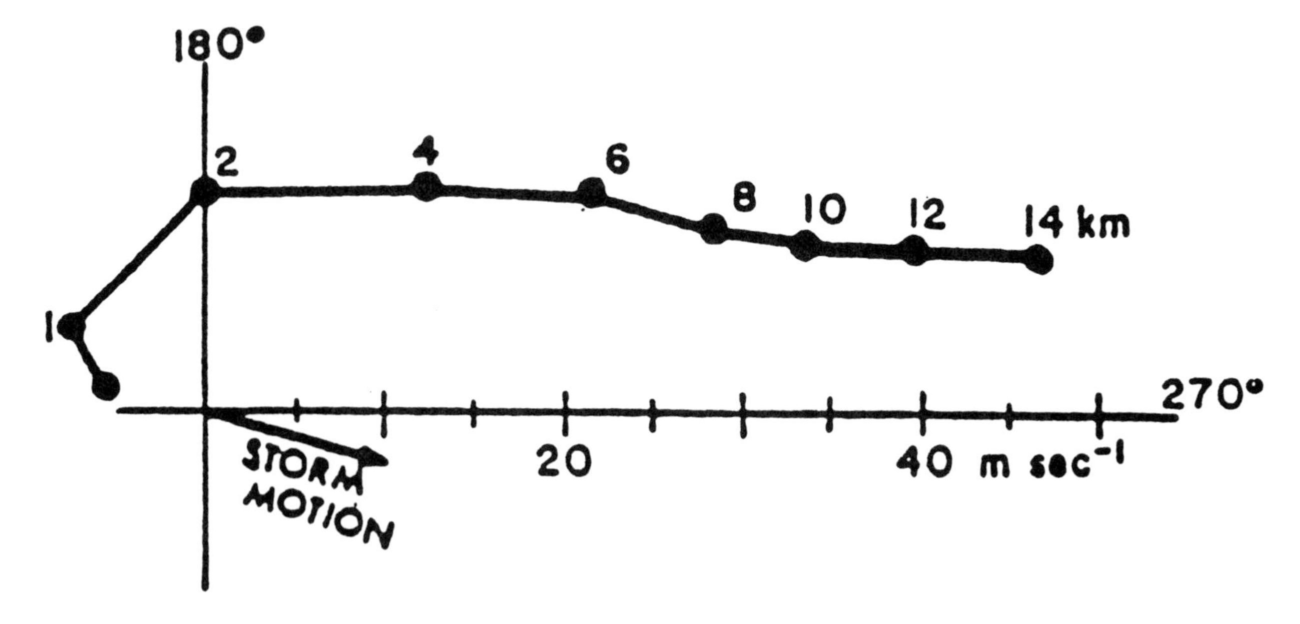

Figure 21-5. A typical hodograph of a Northern Hemisphere supercell environment. Notice the rapid veering of the wind direction in the lowest 2 km (lowest 6000 feet), and then the gradual slight veering with height, all accompanied by a steady increase in the wind speed with height.

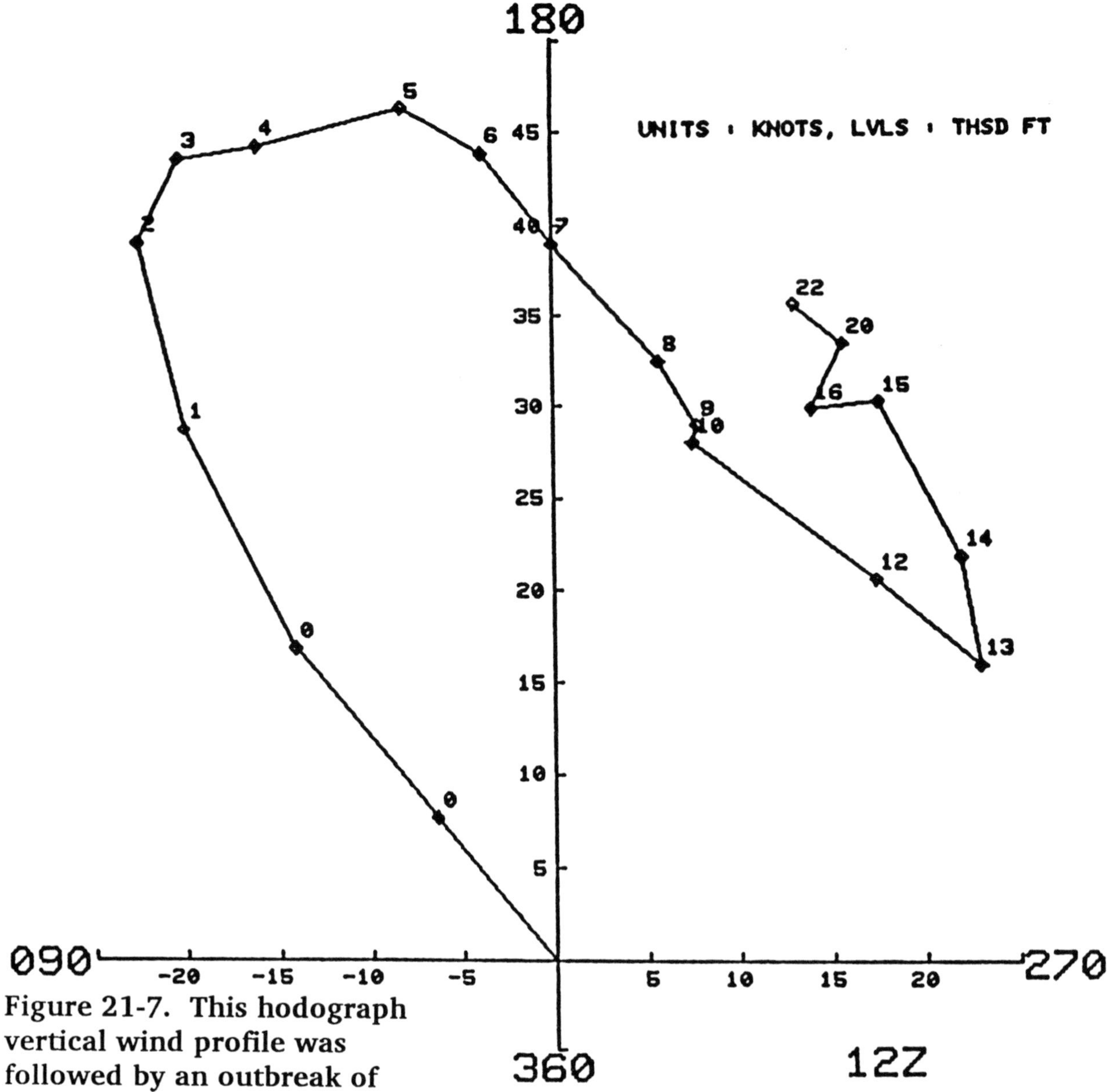

Figure 21-7. This hodograph vertical wind profile was followed by an outbreak of supercell thunderstorms. Here, the wind components are in KNOTS and the end points of the wind vectors are in thousands of feet. The first two indicators, zeroes, are under 1000 feet off the ground. Notice the strong veering of the wind from above the planetary boundary layer (from about 1500 feet) to 13,000 feet. The lifted index was -3, but the air continued to destabilize to lower than -6. The sounding showed two cap inversions, one from 920 mb to 865 mb, an another and weaker one from around 615 mb. Each cap was only about 2 Celsius degrees or less, and each was only about 60 mb thick; thus, they were breakable. The most severe supercells tend to occur when breakable caps exist. These cap inversions temporarily hold back the rising air, but as air mixing ensues, the air surges upwards with rapid severe thunderstorm development. A weak cap in the low-troposphere or mid-troposphere is generally no more than 50 to 60 mb deep and no more than 2 Celsius degrees of warmer temperatures. Caps that are thicker and have greater than 2 C° of warmer temperatures are usually too strong to be readily broken, even if the other three conditions for thunderstorms (lift, instability and moisture) exist.

Since supercells occasionally split into two cells, can the hodograph tell us anything else about these cells? Observations show that if the wind direction from above the planetary boundary layer up to about 12,000 to 18,000 feet off the ground veers with turns (turns clockwise), then the right-moving member of a split pair of supercells tends to become the more severe, and if the wind direction backs with height (turns counterclockwise), then the left-moving storm tends to become the more severe of the two storms.

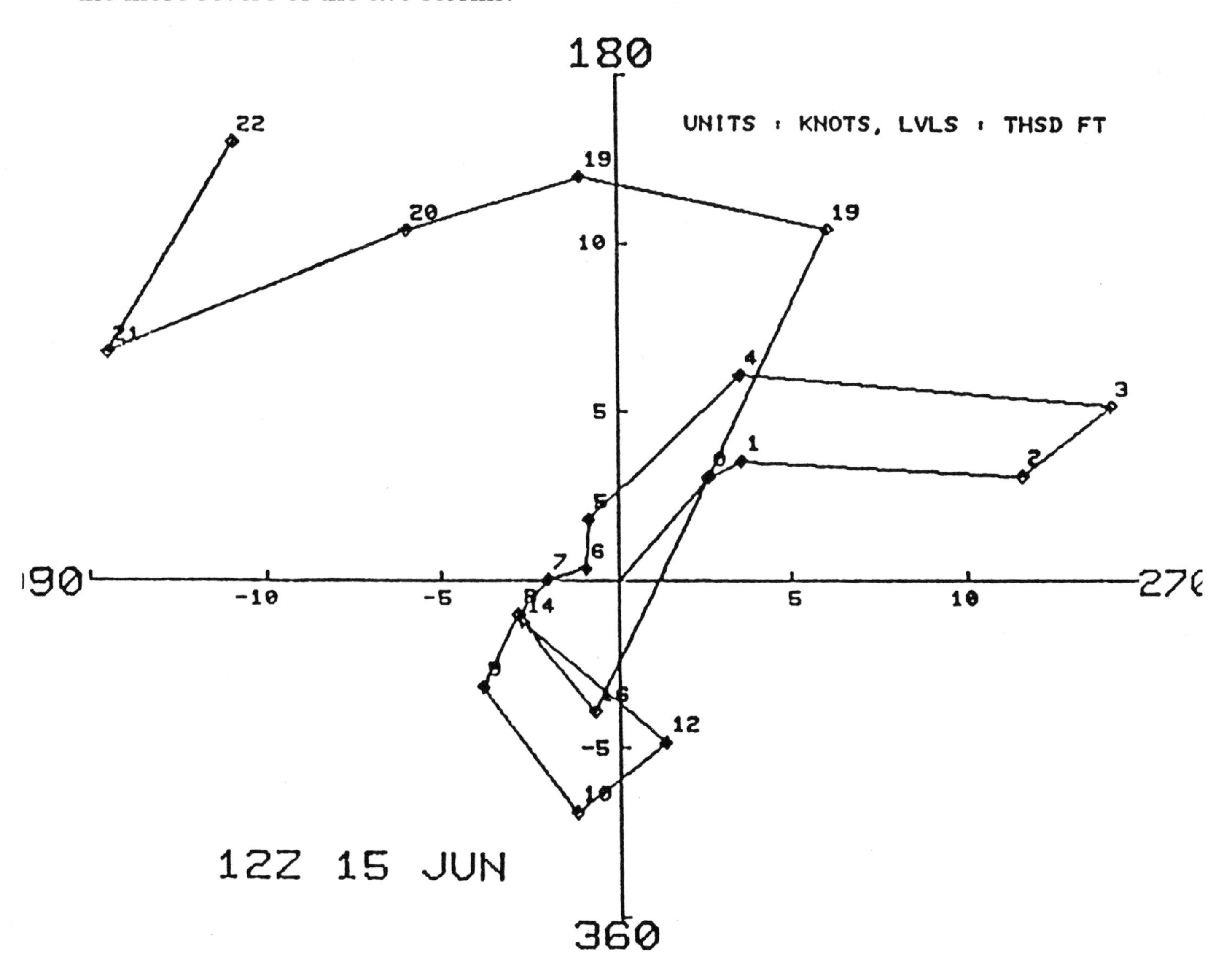

Figure 21-8. This hodograph was followed by an outbreak of pulse thunderstorms, which produced large hail and strong winds. This hodograph goes up to 22,000 feet, showing light and variable winds, essentially below 15 knots, through the layer. But what separates this event from an ordinary thunderstorm is that the lifted index was -5, indicating quite unstable air. Individual intense concentrated updrafts formed, resulting in numerous short-lived convective bursts, the pulse thunderstorms.

Even without strong vertical wind shear, an environment with a very high positively buoyant area (large BRN) can still produce thunderstorms that will give us large hail and downbursts and microbursts. Moreover, when looking at a sounding, if the air is very dry in the low levels but is moist and unstable above that, then if thunderstorms develop they could produce powerful downbursts and the concentrated downbursts known as microbursts. When air is buoyant, all the parcels need is a boost into the buoyant area and then they will accelerate on their own. But if parcels are pushed downward and are colder than the environment, they will also accelerate, but downwards, and can crash into the ground as a downburst or microburst.

Here is an important point: *when the hodograph shows a precursor signature for an environment favorable for a certain thunderstorm type, and the Bulk-Richardson Number is also favorable for that type, then if the four conditions for thunderstorms exist, you are aware of the potential for that type of thunderstorm.*

By way of reference, here is another expression that can be used for the Bulk-Richardson Number:

$$B = g \int_{LFC}^{EL} \frac{T_p - T_e}{T_e} \, dz$$

where T_e is the temperature of the environment, T_p is the temperature of the parcels, g is the acceleration of gravity and z is the height. As you can see by the integral, we are summing up or integrating from the LFC up through the EL.

In summary, using the convective analysis and forecasting tools, we can improve our understanding of and forecasting of thunderstorm development.

Chapter 22. THETA-E: EQUIVALENT POTENTIAL TEMPERATURE

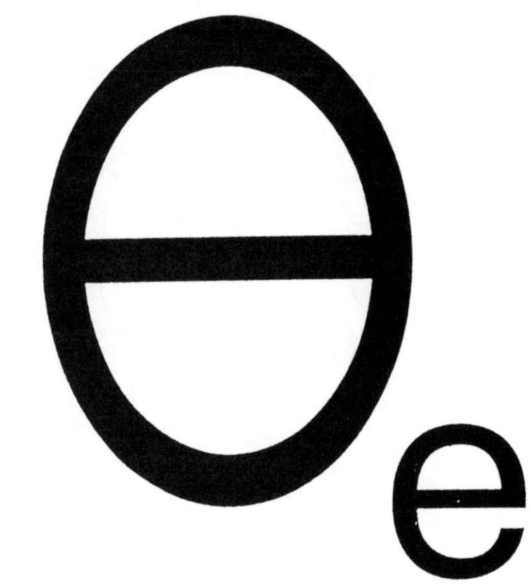

Figure 22-1. The Symbol for Theta-e, the EquivalentPotential Temperature.

One of the most significant breakthroughs in analysis and forecasting has been the evolution of the understanding of how to use the weather parameter known as the equivalent potential temperature, commonly referred to in meteorology as THETA-E.

Theta-e is a method of combining the temperature and moisture content, using the dewpoint for the measure of moisture, into one value, and then plotting these theta-e values on a map, analyzing them in order to find areas of concentrated high values of theta-e, and then using that information to forecast such things as flash flood potential, thundersnow and hurricane movements. Thus, correctly using the analyses and forecasts of the theta-e field are one of the biggest forecast advancements of our era.

Using what forecasters call thermodynamic analysis, parcels of air which originate at a specified level...typically 850 millibars east of the Rocky Mountains and 700 mb from the Rockies westward...are raised until all the moisture is condensed out, and then these parcels are brought back down dry adiabatically to 1000 millibars which is called the potential temperature reference level. Inotherwords, on a thermodynamic diagram or via a computer program, we take the current sounding data from a location, and go through the process of raising air from, say 850 mb, up to a level so high that by then all the moisture is "squeezed out of it" (typically around 200 mb in common computer programs), and then descending the parcel down to 1000 mb. At that point, at 1000 mb, we read the temperature of the parcel. The value it has is its 850 mb theta-e value.

In effect, we are combining the temperature of the parcel with the heat of condensation released as the air becomes saturated and then releases its moisture, and then descending the parcel past its origination point, 850 mb, down to 1000 mb.

The convention is to express the theta-e value in degrees Kelvin rather than degrees Celsius; therefore, add 273 to the Celsius value to get the degrees K value.

Now, the next logical series of questions is, "So what? What good is it? How do we use it?"

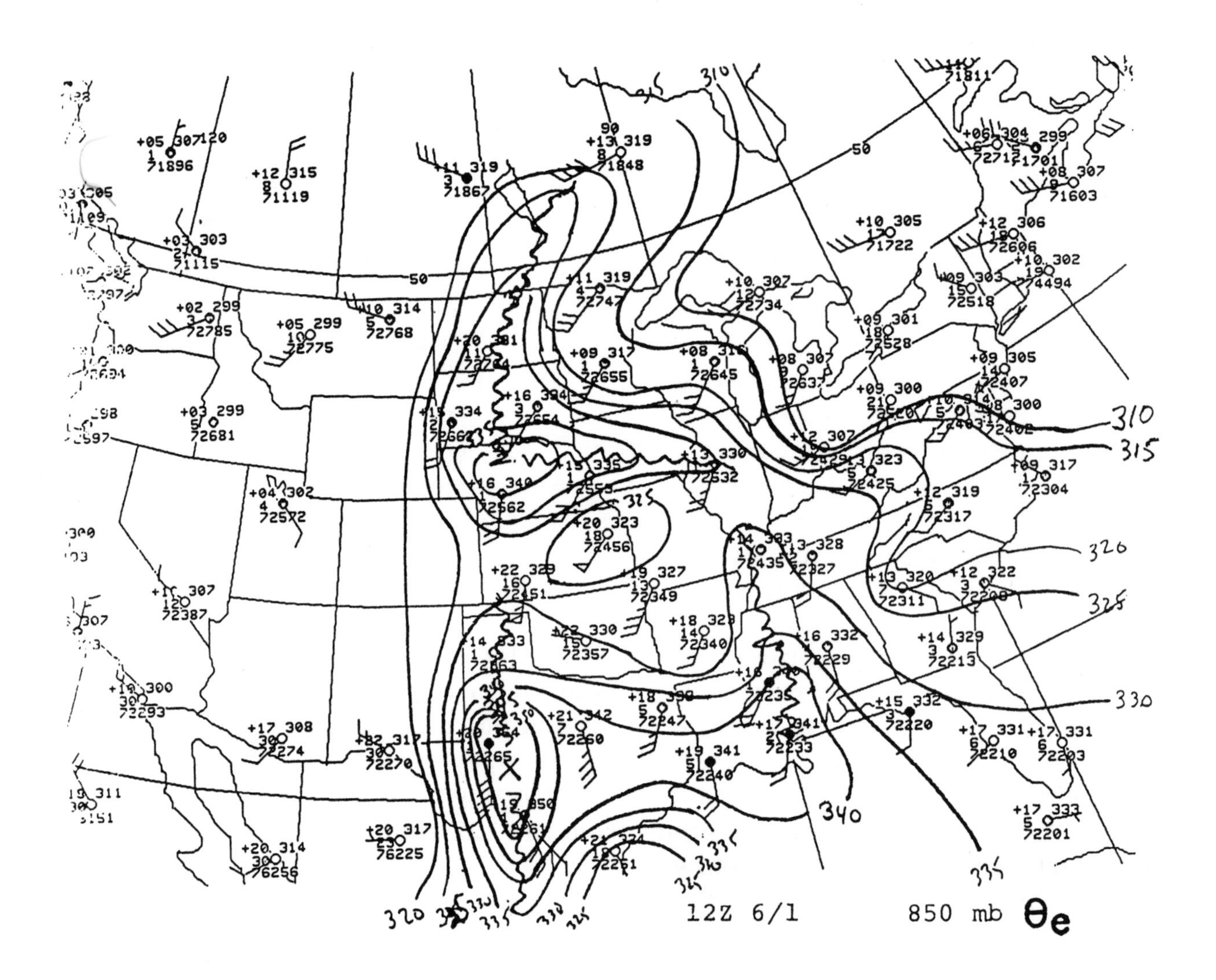

Figure 22-2. An 850 mb Theta-e Analysis. This is an 850 mb chart in which each location's 850 mb height value, in the upper right of each station plot, is replaced by the 850 mb equivalent potential temperature (850 mb theta-e) in degrees Kelvin (°K). For example, the 850 mb theta-e value for Topeka, Kansas is 323 degrees K.

This analysis was done by hand, drawing lines of equal 850 mb theta-e for every 5 K degrees.

850 mb theta-e ridges exist from Nebraska through the Dakotas, from Nebraska through Iowa, in west Texas, in the central Gulf states and a smaller one is observed in Kentucky-West Virginia.

All these ridges were inactive or "wasted theta-e ridges" at this time except for the one extending from the theta-e max in Nebraska east-southeastward through Iowa. In that ridge, some 6 hours later, thunderstorms formed. The surface moisture-flux convergence chart showed some one to two hours ahead of the convective onset where the first cells would likely form.

These cells merged into clusters (with severe weather at some of the merging locales), and the thunderstorm clusters, continuing to feed on high-value theta-e air (which is a concentration of warm, moist air) then merged into a MESOSCALE CONVECTIVE SYSTEM (MCS) by 00Z, about the size of the state of Iowa. An MCS is a large area of organized convection, typically about the size of the state of Iowa, and persisting for 12 hours or more. The MCS that formed on this date produced a flash flood for eastern Nebraska and central and southern Iowa.

This case is typical of how to use theta-e ridges for forecasting organized convection (not single-cell type of convection).

Organized convection did not develop in the other theta-e ridges at this time because: the ridge north from Nebraska was in a region with a strong widespread cap inversion around 700 mb (the region was under a high pressure system with subsidence, i.e, with diverging sinking air), which inhibited convection even though the conditions for convective development existed (low- and/or mid-level lifting mechanism[s], moisture and instability); the air was too stable in the west Texas ridge; the ridge north of the Gulf coast was slightly active in barely unstable air; and there was no lifting mechanism nor instability in the ridge going over West Virginia.

Figure 22-3. Note the theta-e changes from the 12Z chart in figure 22-2 to the 00Z chart twelve hours later, below.

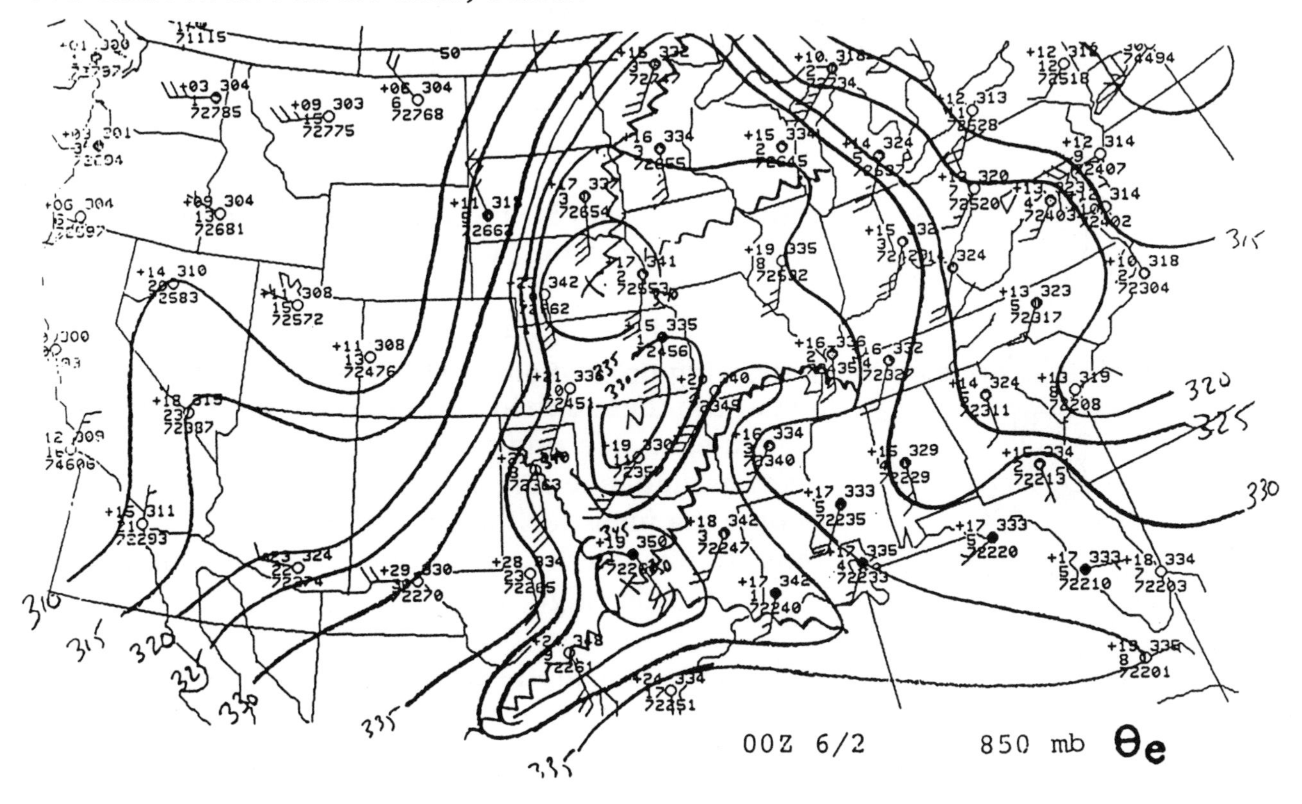

Notice how the theta-e patterns evolve and shift during the 12 hours between the 12Z and the 00Z charts. These changes are gradual. Therefore, because of the conservativeness of theta-e, it can be used to trace the warm moist air it signifies, from chart to chart.

At 00Z, organized convection was underway in Iowa and the theta-e ridge that was over the Dakotas 12 hours earlier had shifted into Minnesota into unstable air without a subsidence inversion aloft, and organized convection began forming within a few hours later over Minnesota in the theta-e ridge.

THE IMPORTANCE OF 850 MB THETA-E RIDGES IN CONVECTIVE FORECASTING: When thunderstorms form in a theta-e ridge, these are the storms that merge and feed off the warm, moist energy supply which is high-value theta-e air (theta-e ridges are concentrations of warm, moist air). These thunderstorms therefore grow into a huge organized mass of thunderstorms and rain known as a mesoscale convective system (MCS). The crucial point about MCSes is that most very heavy rain events (3 inches or more within several hours) are caused by MCSes, so that most flash floods are caused by MCSes.

We now have an analysis and forecasting tool that can be used to anticipate most of the flash floods! This is a major breakthrough in weather forecasting.

A theta-e ridge, when acted upon by a lifting mechanism (e.g., a short-wave trough in the low- and mid-troposphere moving into the theta-e ridge) can be thought of as an axis of available potential energy that can be converted into kinetic energy of the subsequent convection.

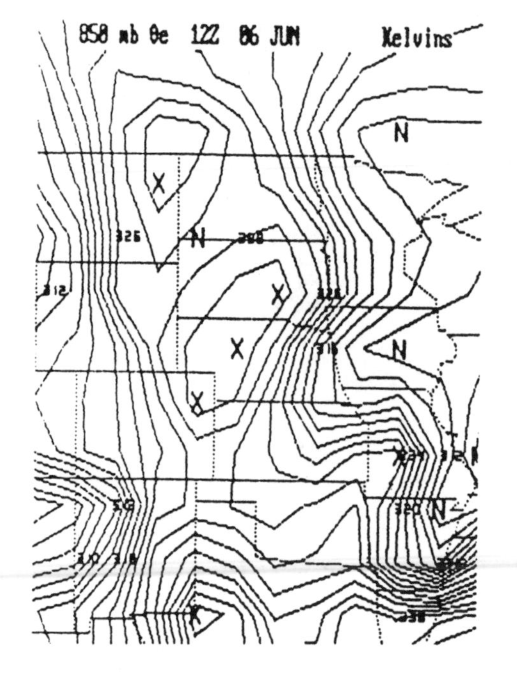

Figure 22-4. An 850 mb theta-e chart showing theta-e analyzed for every 2 degrees.

An analysis in 2-degree increments rather than in 5-degree increments (°K) permits a more detailed look at the theta-e field and helps to find side-lobes of theta-e, or side-ridges. Sometimes even subtle ridges or side-lobe ridges off a main ridge can become the focus for the start of organized convection when the conditions for convection exist within the side-ridge.

An example of a subtle side-ridge in the eastward-poking theta-e ridge into eastern South Dakota.

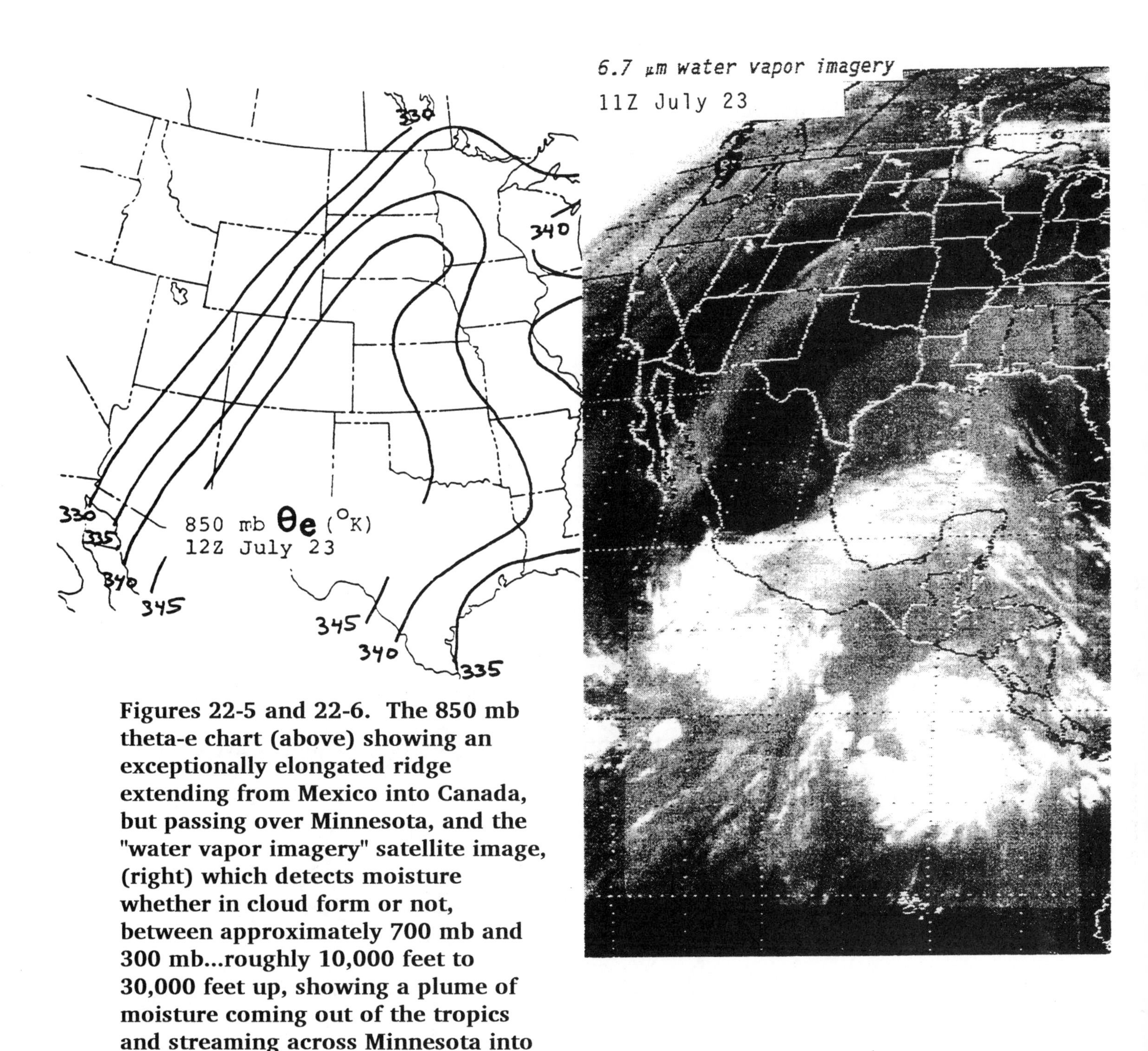

Figures 22-5 and 22-6. The 850 mb theta-e chart (above) showing an exceptionally elongated ridge extending from Mexico into Canada, but passing over Minnesota, and the "water vapor imagery" satellite image, (right) which detects moisture whether in cloud form or not, between approximately 700 mb and 300 mb...roughly 10,000 feet to 30,000 feet up, showing a plume of moisture coming out of the tropics and streaming across Minnesota into Canada.

This plume of moisture is called a TROPICAL CONNECTION. When thunderstorms form in a theta-e ridge, we know that we are likely to experience a mesoscale convective system, which is a very heavy rain producer...over 3" and frequently over 5" of rain...and therefore a flash flood threat, especially if the antecedent local soil conditions are already wet. The flash flood threat is enhanced when a tropical connection exists, with the tropical moisture streaming into the developing MCS.

There will be little moisture added above 500 mb in the troposphere, since the air is too cold to hold copious amounts there, but much moisture will be added from 700 mb to 500 mb, into the developing convection. Moreover, a tropical connection is a continuous infusion of moisture. Therefore, the impact of having a tropical connection streaming into the developing MCS is that it makes the heavy rain producing system an even heavier rain producer. Many of the worst flash floods ever occur with MCSes which form in theta-e ridges and are accompanied by a tropical connection.

In our case study, now look at the 850 mb theta-e analysis twelve hours later:

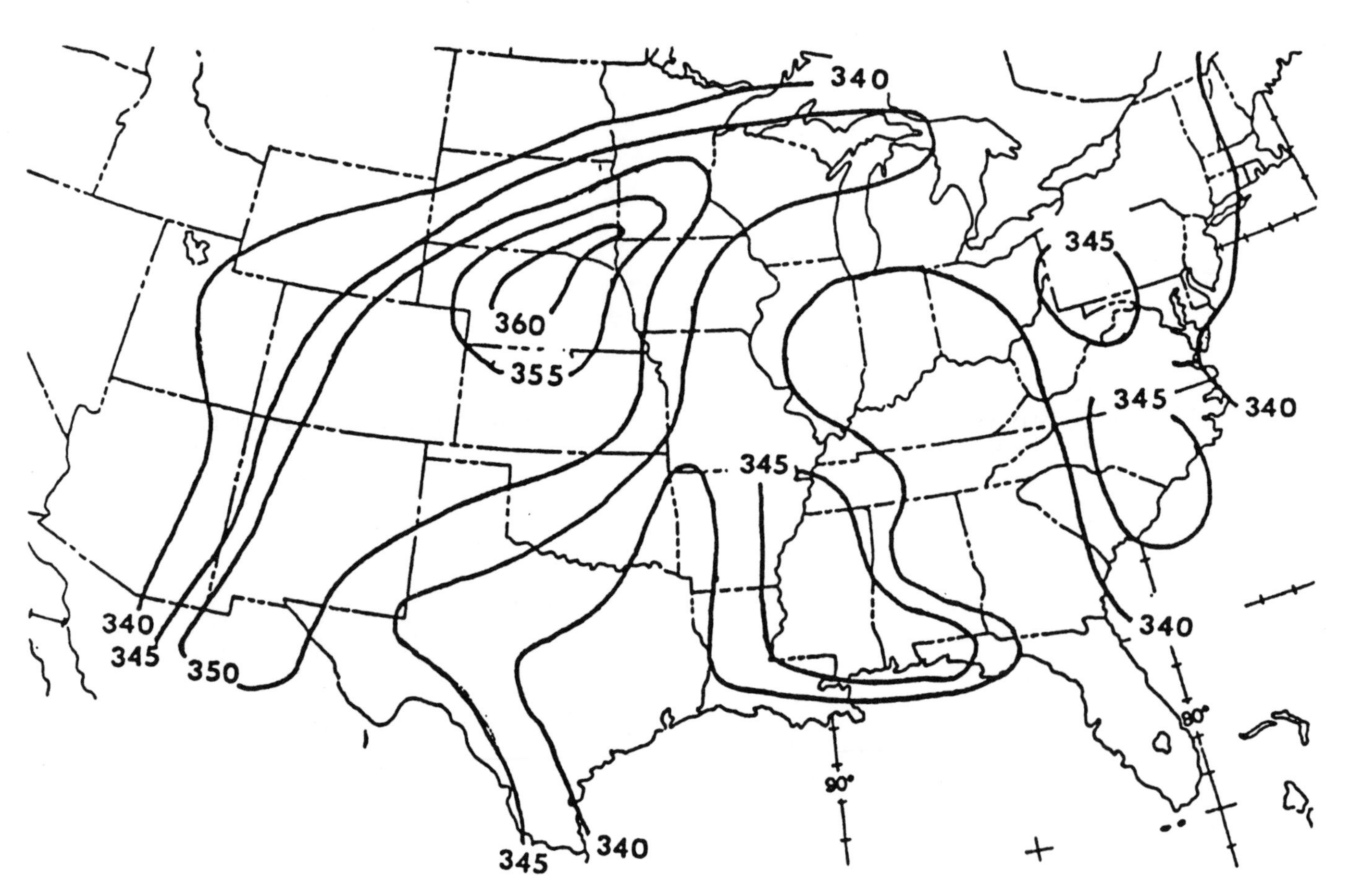

Figure 22-7. The 850 mb theta-e analysis for 00Z on July 24th. Notice how the ridge has intensified from roughly Nebraska into Minnesota. The result was that the convection that was over Minnesota some 12 hours earlier, intensified as it grew into an MCS, and the Minneapolis area received up to 11.1" of flash flooding rainfall.

Important Points About Using Theta-e for Forecasting Organized Convection

• An 850 mb (or 700 mb for western mountainous areas) theta-e ridge or maximum is not necessary to have thunderstorms; a theta-e ridge by itself does not assure thunderstorms; if thunderstorms do develop in a theta-e ridge, then they are likely to merge into clusters with the clusters then merging to form an organized convective system which we call a mesoscale convective system, MCS.

• If in a theta-e ridge and thunderstorms are expected, the surface moisture-flux convergence (SMC) chart typically shows where the first storms will form, namely, in or near a maximum of SMC or in a ridge of SMC, because the chief forcing for thunderstorms is in the lower levels of the troposphere.

• When an MCS is forming in a theta-e ridge, always check the water vapor imagery weather satellite image for a tropical connection into the developing MCS. This plume of warm moist air from the tropics seeds the convective system with additional moisture, resulting in copious rainfall and a flash flood threat.

• Whereas single-cell thunderstorms not in theta-e ridges tend to move with the mean surface-to-500 mb flow, or, for a first approximation, with the 700 mb flow, an MCS in a theta-e ridge typically moves with the 1000-to-500 mb thickness pattern except when the thickness lines diverge, which is called difluent thickness, in which case the MCS remains nearly stationary or propagates backwards.

• The MCS continues as long as the 850 (or 700) mb theta-e ridge is within the thickness pattern; when the thickness pattern (also called the "thermal wind") carries the MCS away from the theta-e ridge, then the MCS starts to die.

Other Uses of Theta-e Analyses and Forecasts

Although using theta-e charts as guidance in forecasting many of the heavy rain/flash flood threats is a major breakthrough in meteorology, various types of theta-e charts are used also to forecast some other weather events:

• heavy rain/snow in western North America

• thundersnow

• fronts aloft

• hurricane movements after landfall

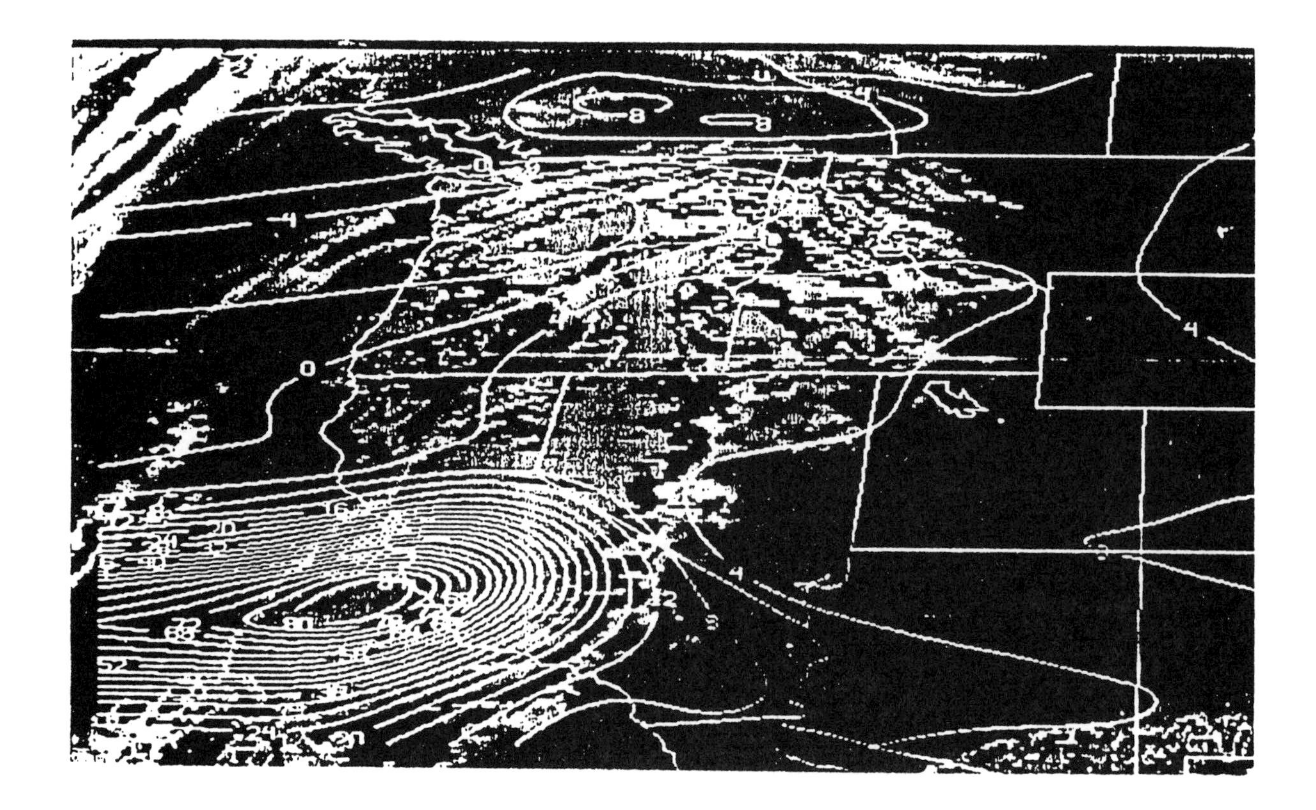

Figure 22-8. A 700 mb theta-e advection chart superimposed on weather satellite imagery.

For west coast heavy rain/heavy snow events caused by a major weather system, especially in the winter and spring, the 700 mb theta-e advection analysis and forecasts give rather good guidance in showing where the heaviest precipitation is likely to fall, namely, within the ridge axis or in the theta-e gradient just north of it. This technique works from central America northward, including Alaska.

An advection chart shows the change in theta-e for a time period, using for twelve hours with this type of usage. Thus, we are using not a theta-e chart but a chart showing the 12-hour increase of theta-e. A decrease of theta-e would represent drier and/or cooler air and would not be conducive to enhancing precipitation. What happens is that the cloud tops tend to grow higher in the advection ridge. There may be thunderstorms embedded in the steady precipitation but this does not necessarily occur. In or just north of this ridge the clouds do, however, show some convective development, growing higher than they otherwise would, producing more precipitation.

Using Theta-e to Forecast Thundersnow

Thundersnow is significant because of its heavy snowfall rates. When a theta-e ridge is superimposed on a synoptic-scale snowstorm, a potential for thundersnow exists. Although warm season convection is quasi-vertical, wintertime convection is often at an angle, and is referred to as SLANTWISE CONVECTION. The cloud tops may be under 20,000 feet high, but the convection may be 25,000 to 45,000 feet long, but occurring at an angle. A useful tool in anticipating possible thundersnow is a THETA-E SOUNDING.

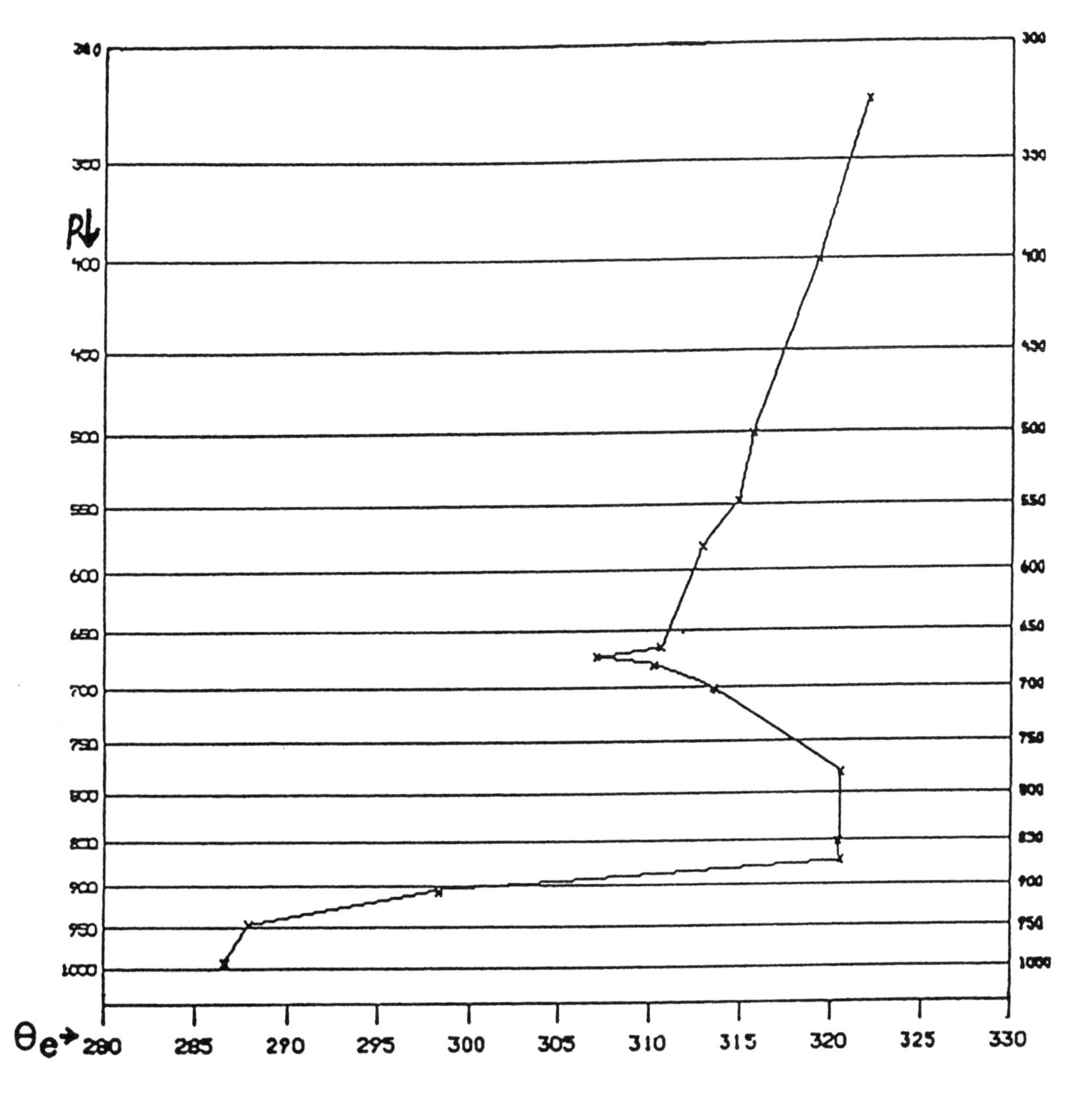

Figure 22-9. A Theta-e Sounding.
The vertical coordinate is millibars or could be the logarithm of millibars, and the horizontal coordinate is the theta-e value in degrees Kelvin.

A theta-e sounding is constructed by taking the air parcel at each level of interest, such as for every 25 or 50 mb, raising the air until the moisture is all condensed out, typically to about 200 mb, and then bringing the air parcel back down dry adiabatically to 1000 mb. For example, to compute the 675 mb theta-e, go to the regular sounding and take the parcel up until all the moisture is condensed out (which we can do graphically on a thermodynamic diagram), then bring it down to 1000 mb at the dry adiabatic lapse rate, and plot its value on the theta-e sounding at the 675 mb level. Connect all the theta-e values for all the levels and analyze the slope of the sounding curve. ***WHERE THETA-E IS DECREASING WITH HEIGHT, THE AIR IS DESTABILIZING.*** If convection were to develop, this is where it would be originating.

The reason why decreasing theta-e with height is what we look for in the troposphere is because there are four ways to destabilize the local atmospheric environment to make the area more conducive to convective development: warm the air in low levels, increase the moisture in low levels, cool the air aloft and dry the air aloft. Any of these or a combination of these allows parcels that have been given a lift to remain warmer than the environment and thus keep rising to form convective clouds and their subsequent showers and/or thunderstorms.

From figure 22-9 we see that the atmosphere is destabilizing from about 775 mb through about 675. Convection, which typically starts at or near the surface in the warmer season, would start aloft, at around 775 mb, which would be about 7,000 to 8,000 feet up. This is an example of ELEVATED CONVECTION. Thus, we can have convection that is both elevated and slantwise.

Now that we know that when theta-e decreases as we rise through the troposphere we are destabilizing that local environment, we now need to determine if there exist a lifting mechanism and sufficient moisture so that we have the conditions necessary for convection: lift, moisture, instability and lack of a strong cap inversion aloft.

If we could generate theta-e soundings across the continent and then look at a cross-section of where decreasing theta-e with height is located and how this region is moving, we would have a guidance tool for anticipating the possibility of thundersnow.

The complex weather graphic on the next page is such a theta-e cross-section.

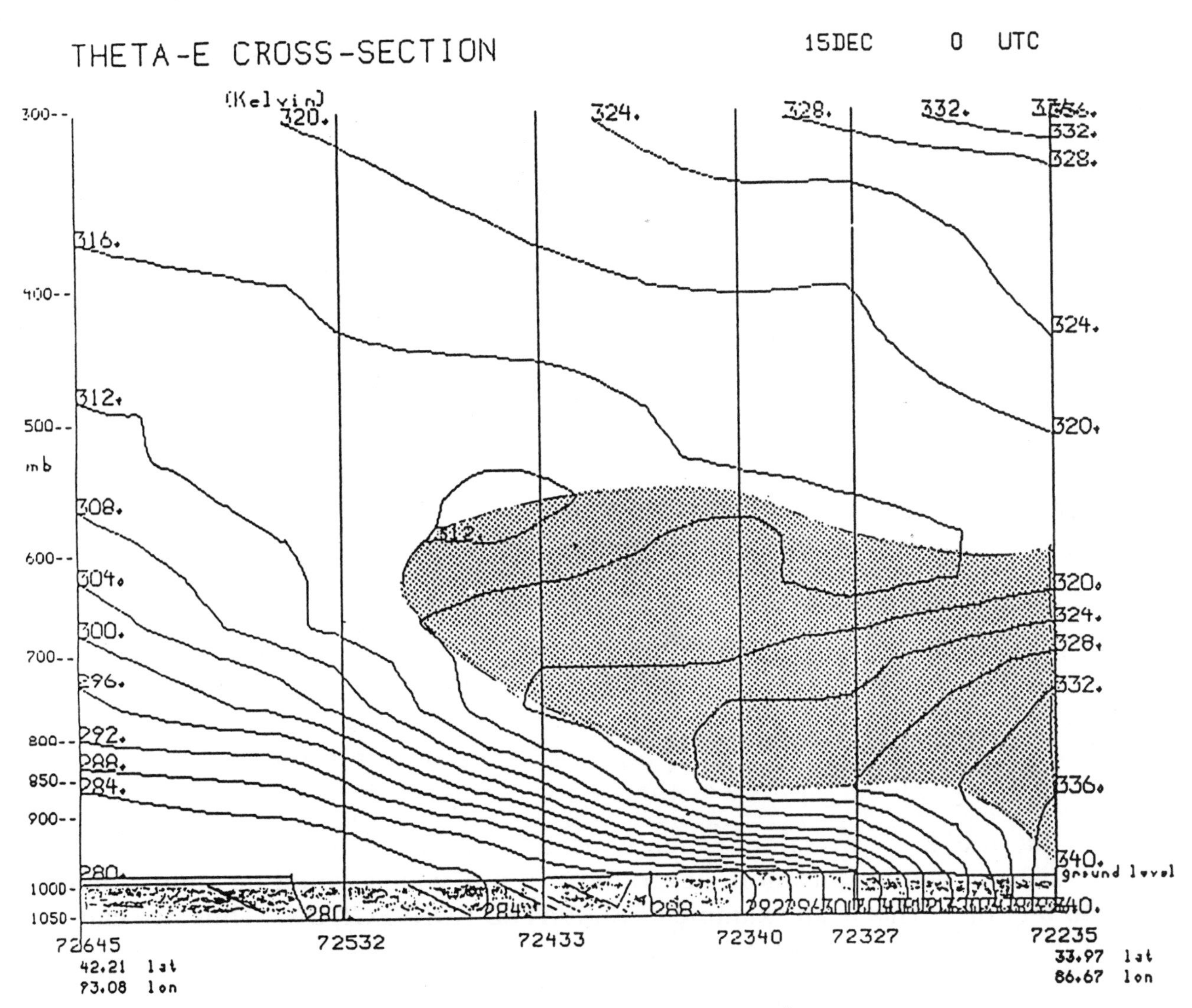

Figure 22-10. A Theta-e Cross-section. The vertical coordinate is pressure, in millibars, and the numbers at the bottom are location identifier for weather stations. Notice at the lower right is the latitude and longitude for a point in the southeastern U.S., and at the lower left is the latitude and longitude for a point in the upper mid-west. Thus, the cross-section runs from the southeastern to upper mid-west region of the country. The station i.d.s, such as 72645 for Green Bay, Wisconsin, are given to identify locations along the cross-section. The theta-e lines from the theta-e soundings are labelled in degrees Kelvin (°K). The shaded-in area is where theta-e values are decreasing with increasing height.

Thus, in this complex type of theta-e weather map, we see by looking at the shaded area, where the potential for slantwise convection exists. The shaded area is moving upwards and towards the mid-west.

In this particular case, a major winter storm was affecting Missouri, and when the "nose" of the shaded area moved into Missouri, thundersnow broke out, resulting in several inches more snow than would have otherwise occurred from the synoptic-scale low pressure system itself. A video-loop of this type of analysis in, e.g., hourly or three-hourly increments would be quite useful to show the progression of this elevated area of instability.

Empirical studies have shown that the decreases of theta-e with height should be at least 5 K degrees, and when they are 10 K degrees or greater, significant convective potential is generated.

Using Theta-e for Fronts Aloft

Although most people are familiar with weather maps showing frontal boundaries on surface weather maps, these fronts also extend for some depth into the troposphere.

It is useful to look for fronts at 850, 700 and even as high as 500 millibars, because such an analysis can be useful in forecasting some types of convection.

When the air gets colder and/or drier at any level in the atmosphere, it lowers the equivalent potential temperature or theta-e for that level. Thus, the front, or leading edge of colder and/or drier air can be found at any level by looking for the leading edge of lower values of theta-e.

Here are two examples of when a front aloft may be a significant factor in convective weather.

Consider an intrusion of cold air flowing over the Rocky Mountains into the Plains. The cold front aloft at about 500 mb would cause cold air advection aloft, which destabilizes the mid- and upper-troposphere, and may lead to convection, or could cause stronger updrafts which lead to more severe convective weather.

Another example occurs when a surface cold front is advancing, trying to dislodge hot air. This is more common in the summertime when the hotter air is more entrenched and the cold fronts are typically weaker than in other seasons. So, the cold front is moving against the hot air but slows down because of the entrenchment and build-up of the hotter air. However, aloft at say 700 mb, the cooler air is still advancing. Ultimately, the front aloft overshoots the front at the surface. The leading edge of this upper front is the leading edge of lower theta-e values. By cooling in mid-levels, this destabilizes the atmosphere and sometimes leads to the creation of a pre-frontal squall-line of thunderstorms. Thus, there are sometimes situations in summertime when a cold front can generate a line of thunderstorms some 100 to 150 miles in advance of its location at the surface, and have another line of thunderstorms along the surface front itself. A theta-e cross-section every one to three hours would therefore be a useful tool for forecasting such thunderstorm potential.

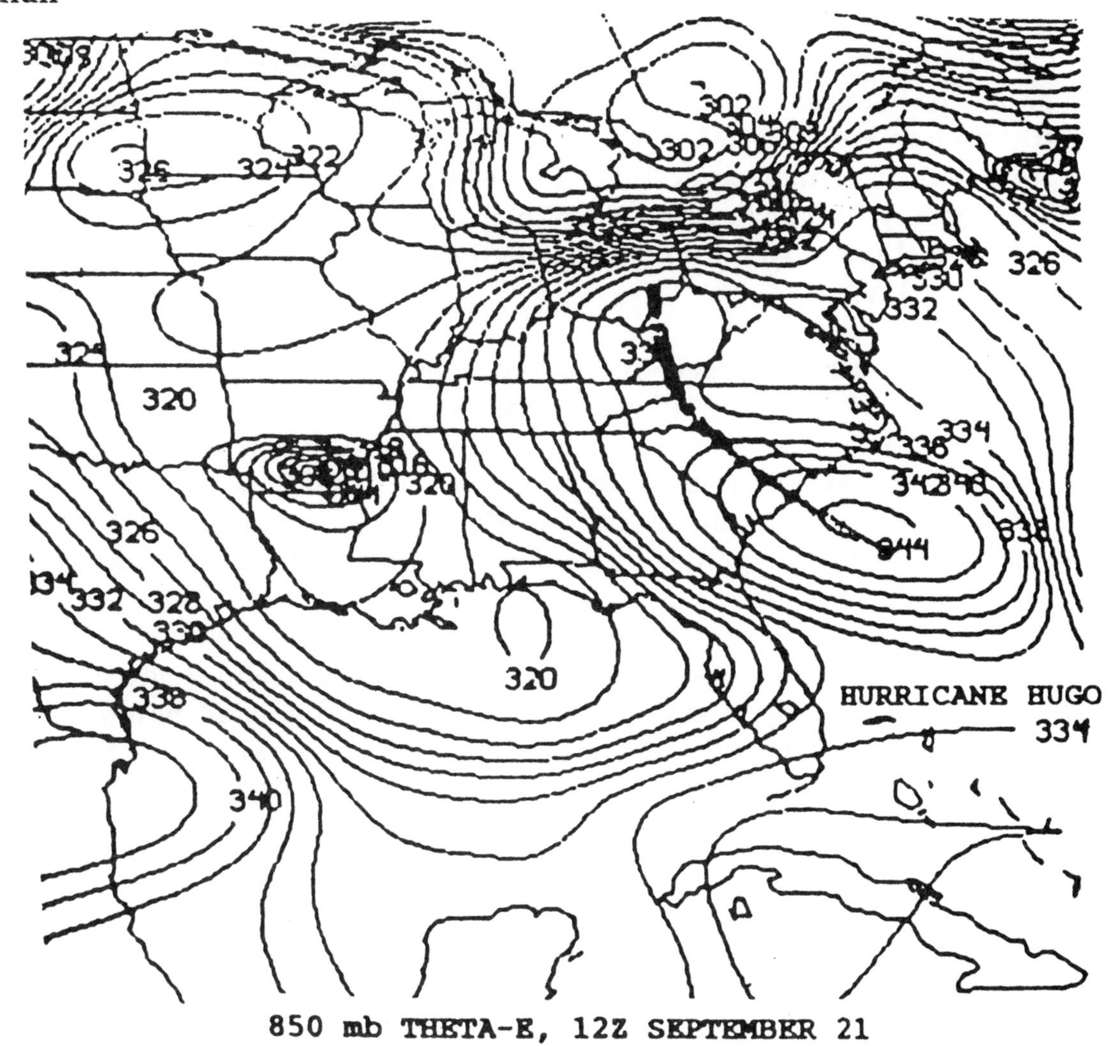

Figure 22-11. The 850 mb theta-e analysis showing a pre-existing theta-e ridge from Charleston, SC to Charleston, WV, and Hurricane Hugo 12 hours before landfall, bringing in its own high theta-e air.

We know that organized convection feeds on concentrations of warm, moist air. These concentrations show up as theta-e maxima and theta-e ridges. Hurricanes are also a type of organized convection; they are a type of mesoscale convective system. We know that they form typically over tropical oceans in areas of high theta-e air.

When these tropical cyclones move inland, they lose access to their chief source of energy, the warm ocean surface and the warm, moist air above it. To maintain their heavy rainmaking ability, they seem to move into a pre-existing theta-e ridge over land, at least initially. As they move into higher latitudes, the strong westerlies...winds aloft...tend to take over as the chief steering influence for their movements.

In the Hurricane Hugo case in figure 22-11, note the theta-e maximum off the South Carolina coast. This represents the eye or center of the hurricane. Hurricane Hugo is moving inland, transporting its own high theta-e tropical air. This theta-e air must be distinguished from any pre-existing theta-e ridges near the landfall area. In this case, a pre-existing 850 mb theta-e ridge was in place from South Carolina into West Virginia. Although all the computer weather forecast models predicted that Hugo would hit the mid-Atlantic coast and then turn sharply to the right, striking New Jersey, the New York City area, Long Island and New England, instead Hugo moved into and through the theta-e ridge from Charleston, SC to Charleston, WV.

The theta-e analysis has been used since Hugo, typically showing hurricanes moving into pre-existing theta-e ridges. A notable hurricane-theta-e interaction was noted in 1992 when Hurricane Andrew crossed southern Florida, travelled through the Gulf of Mexico and then made its second landfall, striking the Louisiana coast. After hitting the coast, Andrew ran into a theta-e TROUGH, i.e., a concentration of low theta-e value air. Andrew abruptly veered rightward directly into a theta-e ridge.

In conclusion, the 850 mb theta-e analysis appears to be a useful guidance product for how a hurricane is likely to move during the approximately 24-hours after making landfall: **if a pre-existing theta-e ridge is in the proximity of the landfalling hurricane, then the storm is likely to move into the ridge initially.**

El Nino and Theta-e

El Nino is a sudden warming of a vast area of equatorial Pacific Ocean surface (and for at least some depth below the surface) water, in the middle of the Northern Hemisphere Pacific Ocean. El Nino may start in Southern Hemisphere waters off Peru and rapidly work its way northwestward into the south-central Northern Pacific Ocean. El Nino gets its name for the Christ Child, since it typically starts about November and peaks in December through March. (El Nino means little baby boy.)

El Nino does not occur every year, and very strong El Ninos are usually several years apart. The El Nino event of 1992-1993 was especially interesting, because it persisted through the summer of 1993 and appears to have played a role in the disastrous mid-west floods of that year, because it created a tropical connection that persisted through the summer, with that continuous mid- and upper-tropospheric moisture injection feeding into persistent 850 mb theta-e ridges over the mid-west United States.

The cause of El Nino is unknown. Some type of dynamic heating process is suspected. Thus, El Nino remains a fascinating mystery of oceanography.

When such a vast ocean surface warms up...sometimes by more than 5 Fahrenheit degrees, it warms the air above it. Warmer air can hold more moisture than when it was cooler; consequently, this warm air absorbs more water vapor from the ocean. Much of this moisture works its way up to mid-levels of the troposphere (700 to 500 mb), and some of the moisture is transported to the upper-troposphere (above 500 mb, up to 300 to 200 mb).

Next, we would want to know how the moisture gets transported across the Pacific into North America in tropical plumes. We know that tropical connections occur throughout the year around the globe, but when El Nino occurs, we have intense and persistent tropical connections, with major, continuous infusions of moisture from the tropical North Pacific into North America in the mid- and upper-troposphere.

There are two chief sources of this transport: the sub-tropical jet-stream and anticyclonic outflow aloft from organized convection in the tropics.

Jet-streaks of the sub-tropical jet readily carry the air and its moisture from the source region for this tropical moisture east-northeastward across the central and eastern Pacific and over the North American continent.

The other major source of this moisture plume is in the Intertropical Convergence Zone (ITCZ). There are typically large high pressure systems over the North Atlantic and North Pacific Oceans, and there are also large highs over the South Atlantic and South Pacific Oceans. The low pressure systems move through and around these highs. The circulation around a high is clockwise in the Northern Hemisphere and counterclockwise in the Southern Hemisphere. This results in air coming together or converging near the equator, with climatological statistics showing us that the greatest convergence is a few degrees latitude north of the equator. This is the axis of the Intertropical Convergence Zone. The ITCZ axis migrates some to the north and south, but stays north of the equator. Most hurricanes form in the ITCZ since it is a zone of converging, rising and very warm tropical air...thus, it is air of a high-value theta-e environment. Hurricanes form out of organizing convection which is an MCS, mesoscale convective system. Since air is converging and rising into the MCS in low and middle levels of the troposphere, the air must come out of the system and diverge aloft. It does so as anticyclonically-curved plumes of air, which also contain the moisture.

Thus, each of these MCSes in the tropics, including those that become hurricanes, generate their own tropical moisture plumes.

These MCSes form in regions of high values of low-level theta-e.

In conclusion, when El Nino occurs, the tropical connections are more intense and prolonged, and if they stream into MCSes over North America forming in theta-e ridges, the heavy rain/flash flood
potential from these MCSes is enhanced.

MORE SPECIFICS ON USING THETA-E FOR PREDICTING HURRICANE MOVEMENTS:

Because more weather observations are available over land than over the ocean, it is easier to generate theta-e analyses over land than over ocean areas. Theta-e may be useful in hurricane forecasting as following:

- Separate the 850 mb theta-e envelope that moves with the hurricane from any pre-existing 850 mb theta-e ridge over land. As the hurricane approaches land, it tends to move into the theta-e ridge and maintains its heavy rain producing factory even when it has been downgraded into a tropical depression low-pressure system. Although the hurricane and then its remnants tend to move into a pre-existing theta-e ridge, the system will likely be steered more by the mid- and upper-level winds as it moves into higher latitudes, especially from about 40°N or higher.

- In the absence of a pre-existing and well-defined 850 mb theta-e ridge over land near where the hurricane is heading towards, the hurricane's own 850 mb theta-e ridge is usually still useful in short-term forecasting. The ridge typically "pokes" in the direction towards which the system is moving. Thus, analyzing the storm's own 850 mb theta-e envelope of warm, moist air shows how the storm is likely to move for up to 12 to 18 hours.

The initial history of using the 850 mb theta-e analysis with a landfalling tropical cyclone shows the following:

- In 1989, Hurricane Hugo, which was predicted by the computer forecast models to hit the Carolinas and then move up the east cost into the New York City area and New England, instead moved into a pre-existing theta-e ridge that stretched from Charleston, South Carolina to Charleston, West Virginia. Hugo moved from Charleston, S.C. to Charleston, W.V.

- In 1990, Klaus and Marco merged, moving out of the eastern Gulf of Mexico as Marco. Although the models had the storm center moving east of the Appalachians across eastern Georgia, Marco went directly into an 850 mb theta-e ridge west of the mountains and gave portions of Georgia significant flash flooding.

- In 1991, Hurricane Bob travelled directly up a theta-e ridge off the east coast, directly into New England.

These were the only tropical storms to strike the mainland within that time-period, and was the first period of time that the 850 mb theta-e charts were used to study the movements of these tropical cyclones as they were making and after they made landfall. Since the theta-e analysis proved helpful in these initial study-cases, some forecasters have continued experimenting with its use for landfalling tropical systems.

For example, in 1995, the path of Hurricane Erin was easily predictable through using 850 theta-e from the Gulf states into New England. The heavy rains and flash flooding occurred as Erin continued to feed on high-value theta-e air (i.e., on a concentration of warm, moist air), as she moved through a pre-existing theta-e ridge.

MCSes forming over land areas in low latitudes and moving out over the ocean can become tropical cyclones. The MCSes can originate over Texas, Africa and Asia, for example. MCSes form only in theta-e ridges. Unlike air mass or single-cell thunderstorms, most of whose life-spans average from about 20 to 60 minutes, a mesoscale convective system persists for many hours, in many cases over 9 hours. Therefore, an MCS needs an energy source to sustain it, which is the concentration of warm and moist air which comprises the theta-e ridge.

When an MCS moves out from the land to over the ocean and the conditions there, as outlined in earlier chapters of this book, are favorable for tropical cyclogenesis, then if the MCS holds together sufficiently long, it may evolve into a tropical storm and subsequently likely reach hurricane status.

Indeed, if you study closely the enhanced infra-red weather satellite images over equatorial Africa during the hurricane season, you observe that most of the hurricane seedlings that move into the Atlantic as easterly waves originate as mesoscale convective systems.

Moreover, as a tropical cyclone forms in the Gulf of Mexico, Caribbean Sea, the rest of the low-latitude North Atlantic Basin, the low-latitude Pacific Basins and the Indian Ocean, it mushrooms as an MCS before growing into a tropical storm, in most of the tropical cyclogenesis episodes.

The relationship of concentrated high-value theta-e air and tropical cyclones is therefore established.

The following is from research done by severe weather forecasters of the U. S.'s National Weather Service, and applies to the Northern Hemisphere. The places labeled "SVR AREA" are areas of greatest severe weather potential.

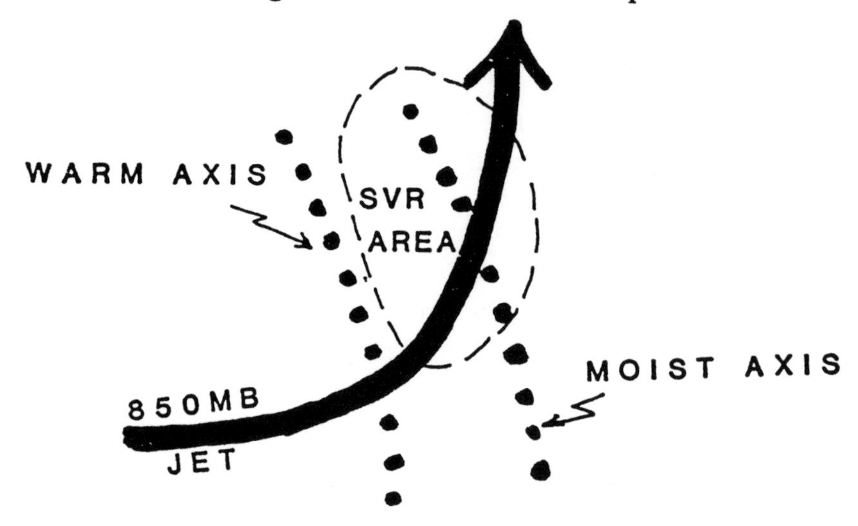

Figure 23-1. A favorable pattern for severe weather is where the low-level warm axis is upstream of the low-level moist axis. The warm axis is usually southwest or west of the moist axis.

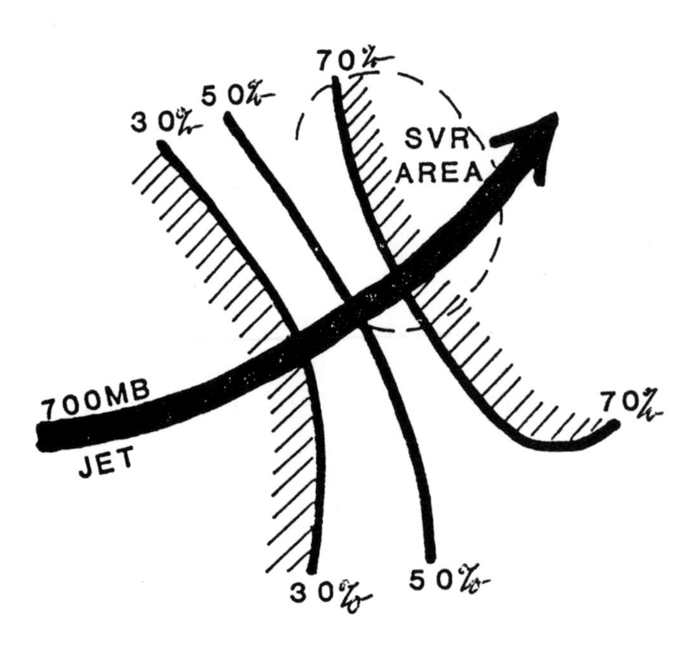

Figure 23-2. A favorable severe weather pattern is when the 1000-to-500 mb mean relative humidity gradient is large, with the 30% isoline to the southwest, and with a 700 mb jet crossing the contours as depicted. Thunderstorms are typically found between the 45% and 75% isolines.

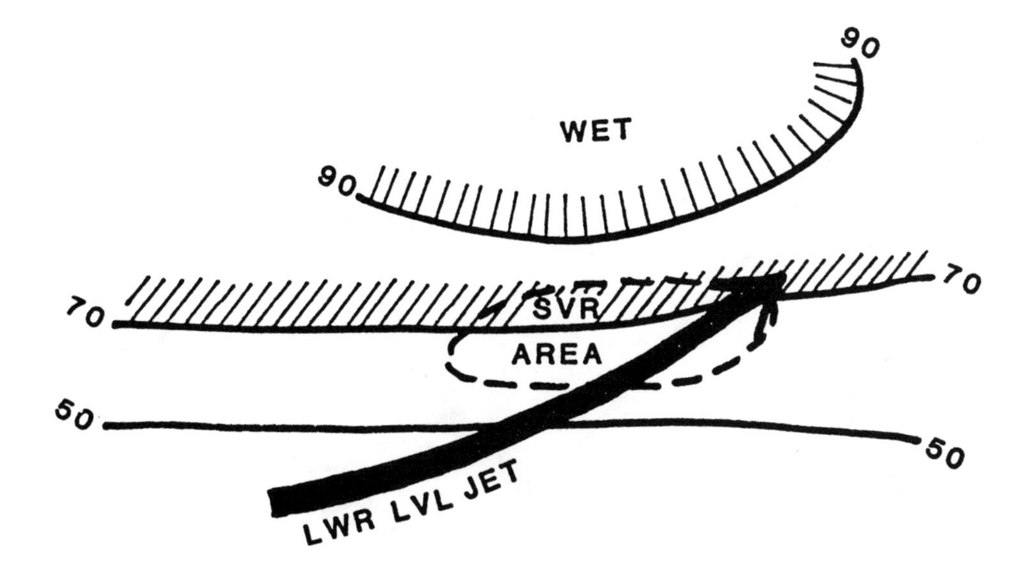

Figure 23-3. Severe thunderstorms sometimes develop when the 1000-500 mb mean relative humidity pattern is as or like the pattern depicted above. This condition is often associated with a weak short-wave trough and strong winds.

Figure 23-4. The greatest threat for severe thunderstorms with this 850 mb pattern is usually along and to the left of the low-level jet, downstream from a wind maximum.

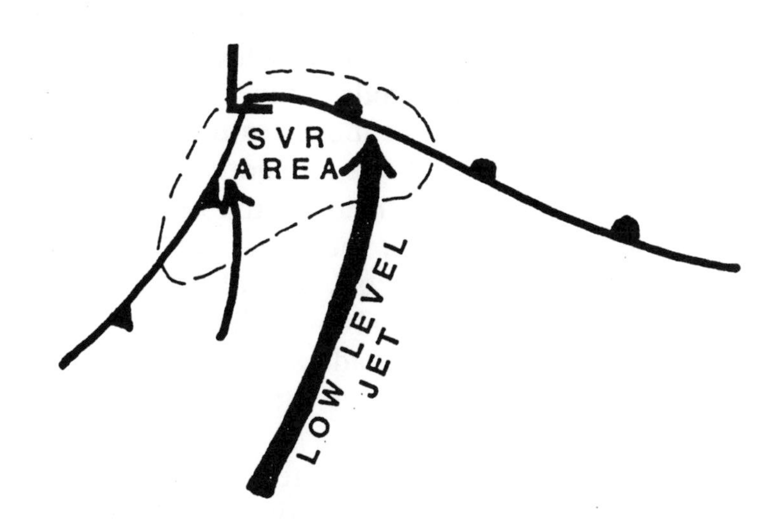

Figure 23-5. The severe thunderstorm threat is enhanced when the low-level jet acts across a boundary where convergence is already occurring.

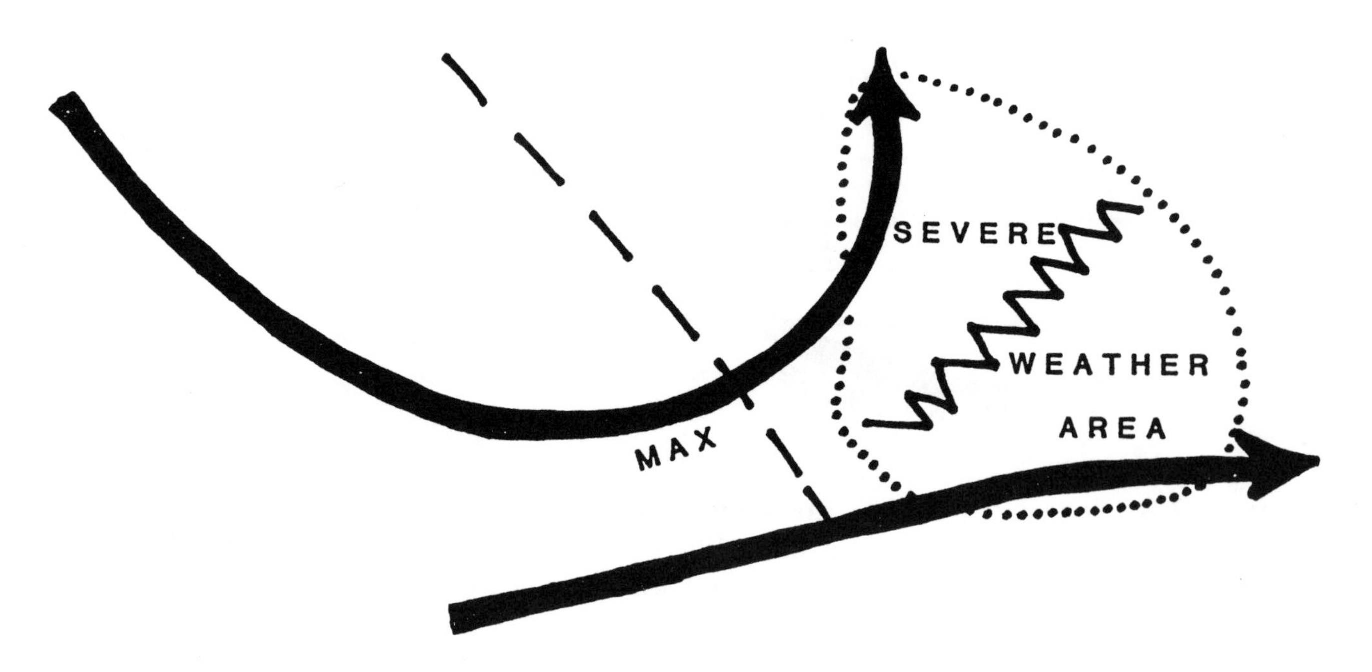

Figure 23-6. The solid dark curves are the 500 mb wind flow. The trough is at 850 mb. This pattern is associated with major severe weather outbreaks during the convective season. This, or a similar pattern, usually occurs when a strongly digging short-wave moves into a profoundly difluent region aloft, in which the difluent area is above a negatively-tilted (northeast-to-southwest oriented) trough. [For the nonmeteorologist unfamiliar with these terms, we recommend your obtaining a copy of the book, "WEATHER MAPS - How to Read and Interpret all the Basic Weather Charts", available for $29 from Chaston Scientific, Inc.; P.O. Box 758; Kearney, MO 64060. Thousands of copies of this practical book have been sold in the U.S., Canada and Europe and it used by colleges and universities. The "Weather Maps" books is the standard reference on the topic.]

The severe weather threat area is typically ahead of the short-wave axis as it moves into the region just described, as depicted on the sketch.

This pattern is typically associated with a significant surface low pressure system. If the air is widely unstable and contains copious moisture (high dewpoints), then the severe weather outbreak can occur over a large area.

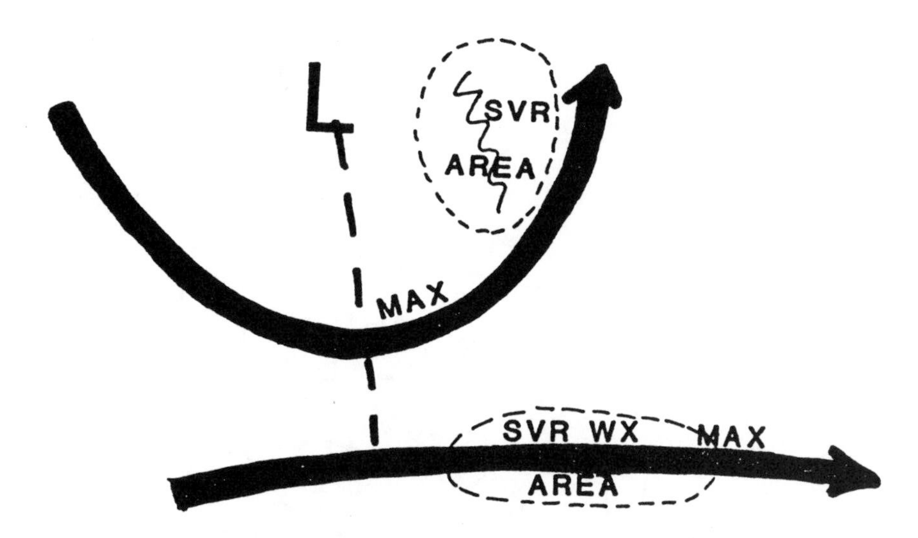

Figure 23-7. When the above 500 mb pattern is accompanied by a large amount of low-level moisture near the southern branch of the 500 mb "jet", then only one severe weather area may occur; it would be the southern area, as depicted. However, when marginal moisture exists farther north, then a second severe wether threat area may occur near the surface low, inthe region of strongest positive vorticity advection.

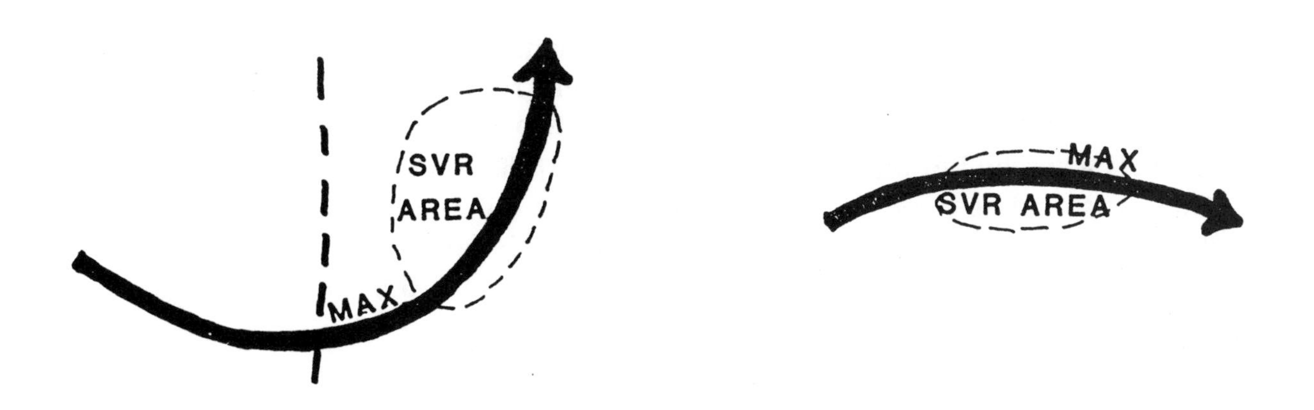

Figure 23-8. When the "ingredients" for convection exist, the summertime (about 200 mb) jet-stream jet-streaks can, by enhancing upward motion in their upper divergent areas, enhance thunderstorm development. The left-front and right-rear quadrants of a fairly straight jet-streak are favored locations. With a cyclonically-curved jet-streak, the upper divergence tends to spread out more north of the jet axis, and with an anticyclonically-curved jet-streak, the upper divergence tends to spread out more south of the jet axis. Since upper divergence is above the level of nondivergence (LND) (about 550 millibars [hecto-Pascals]), then convergence is occurring below the LND, and upward motion is occurring through the layer.

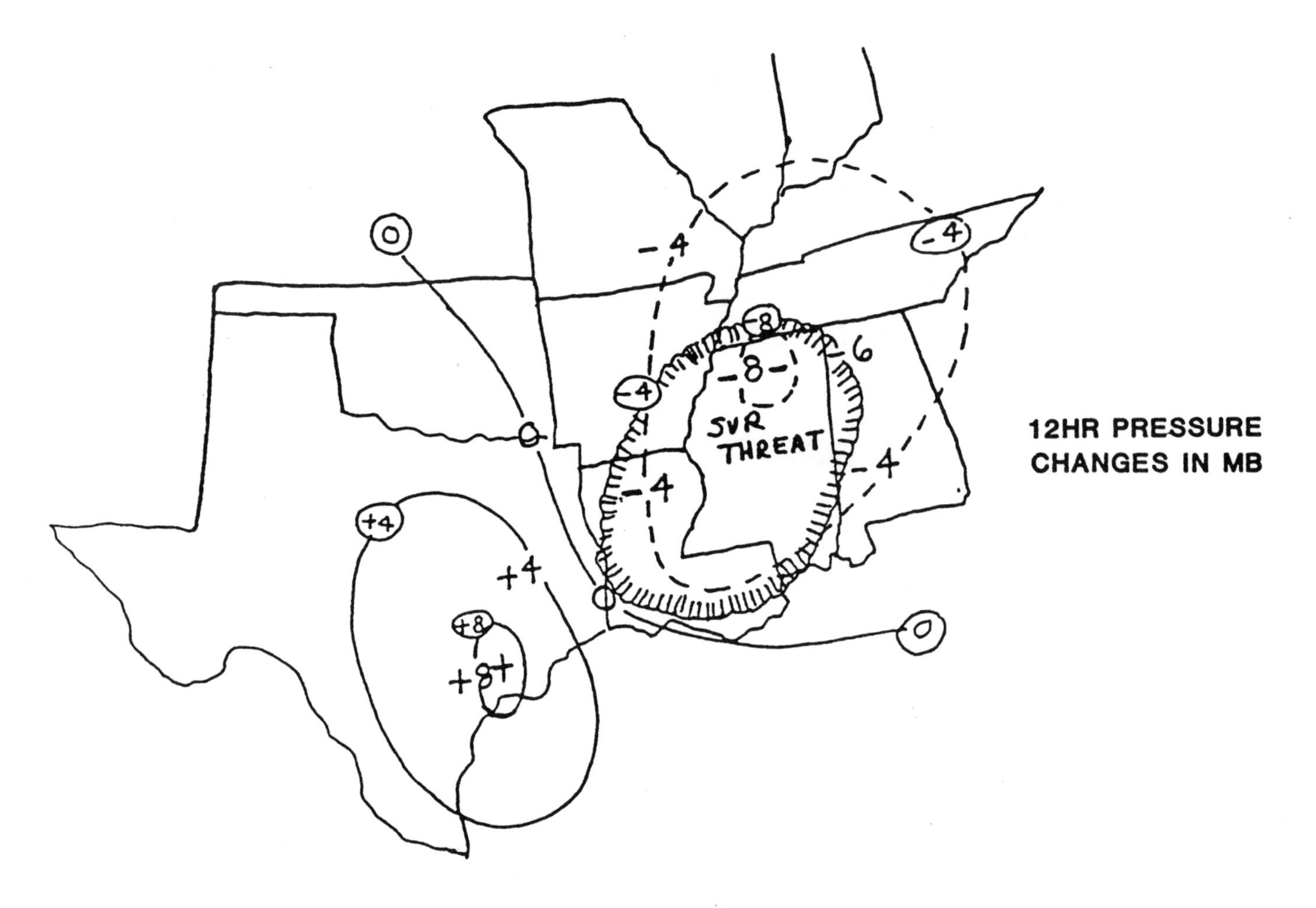

Figure 23-9. Organized falling surface pressures can be a precursor of organized severe thunderstorm activity. The severe potential is enhanced when a complementary pressure rise occurs, moving toward the area ρf falls. In general, the larger the magnitude of the fall-rise couplet, the greater the potential for severe thunderstorms.

In a region that has the conditions for convection, often the analysis of the surface moisture-flux convergence can pinpoint where the first storms are likely to form, usually one to two to three hours before the storms form. A bullseye of surface moisture-flux convergence shows where the low-level winds are transporting in air that also contains moisture. Physically, this chart shows where air is converging and moisture is building up. [Again, please refer to the book, "WEATHER MAPS - How to Read and Interpret all the Basic Weather Charts" for explanations on how to read and use such charts.]

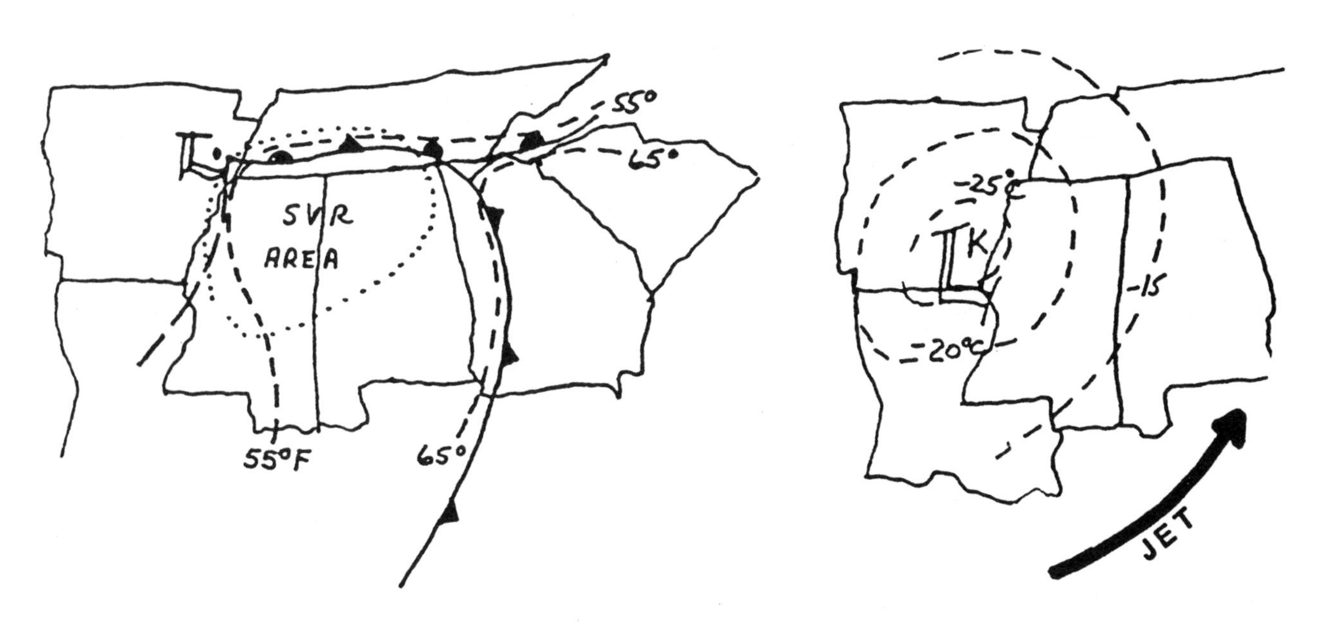

Figure 23-10. A cool pool aloft, which may be associated with a cut-off (closed) low, destabilizes the troposphere aloft and as it moves over an areas of moist air, severe thunderstorms, which can produce hail, may develop.

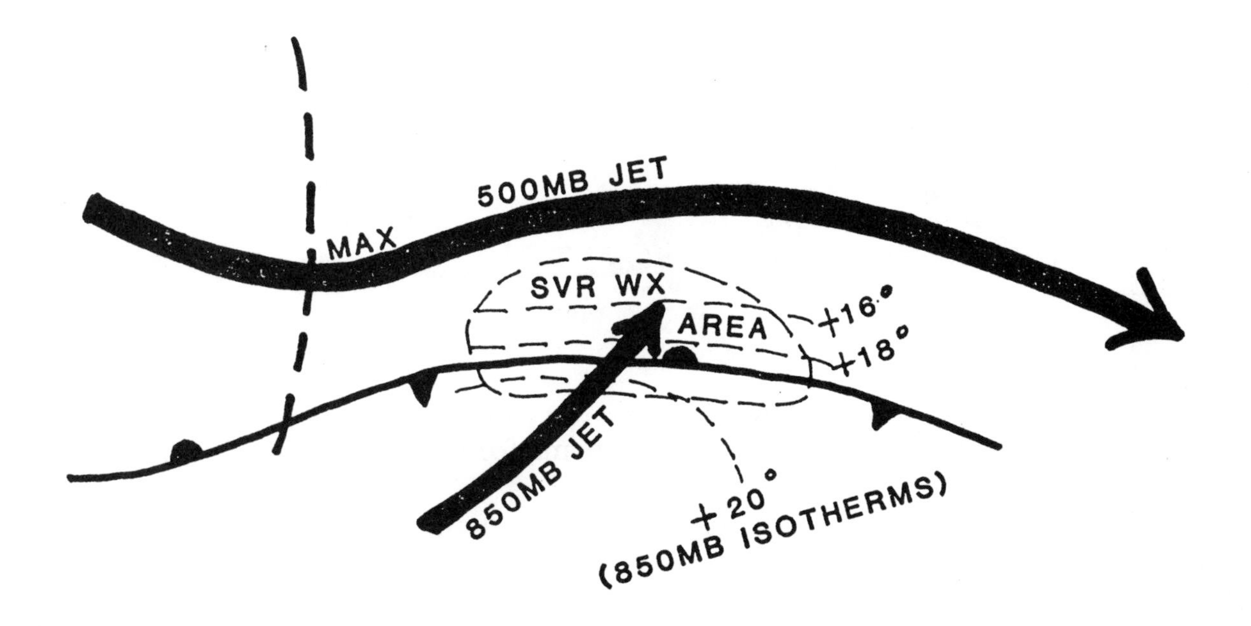

Figure 23-11. When widespread air mas instability is occurring, the upper dynamic features enhance the lower contributing features such as low-level warm air advection. The above sketch is an example. The severe weather threat area here is south of the 500 mb "jet' (or use the 200 mb jet in summer), in the northern region of the low-level warm air advection.

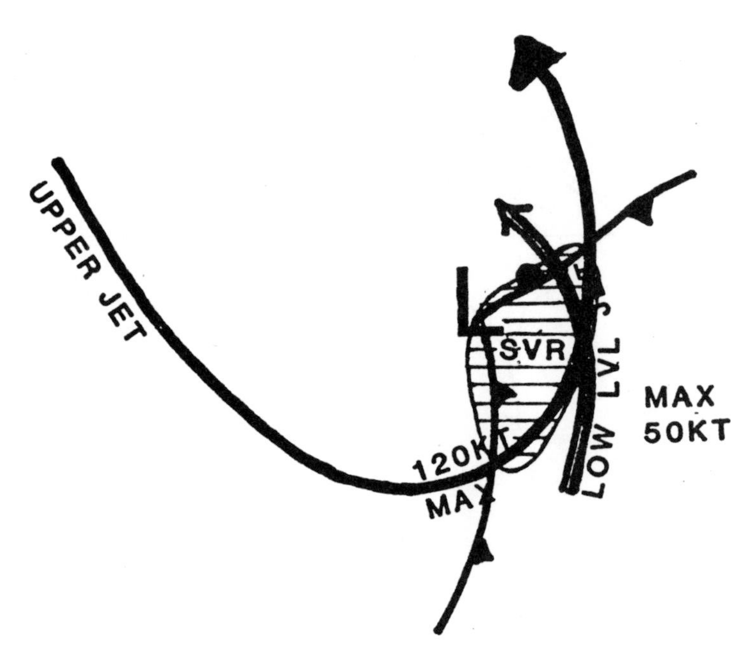

Figure 23-12. In the winter and in the early spring, in episodes of marginal to moderate instability, the axis of severe weather potential is often closely associated with an upper divergent pattern.

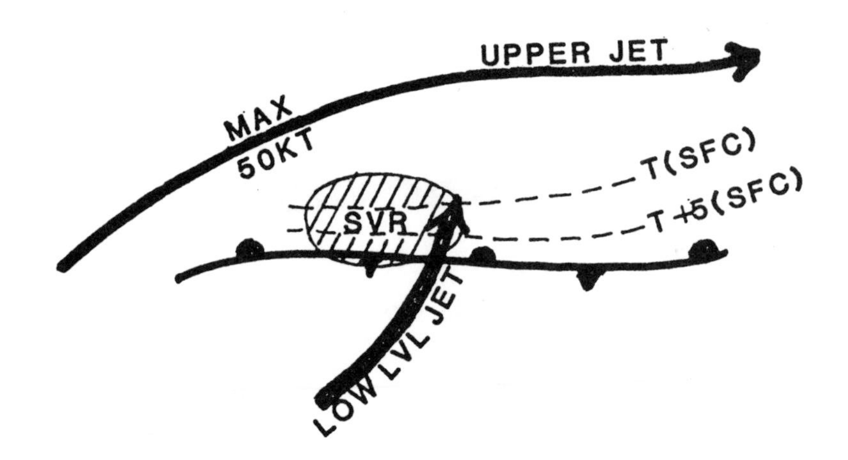

Figure 23-13. In the summertime, with very unstable to extremely unstable air, the severe weather potential is often closely associated with the low-level warm air advection.

Chapter 24. RADAR SIGNATURES OF TORNADOES

Tornadoes are detected on radar by various signatures on the reflectivity displays (precipitation displays), and on the Doppler wind velocity displays.

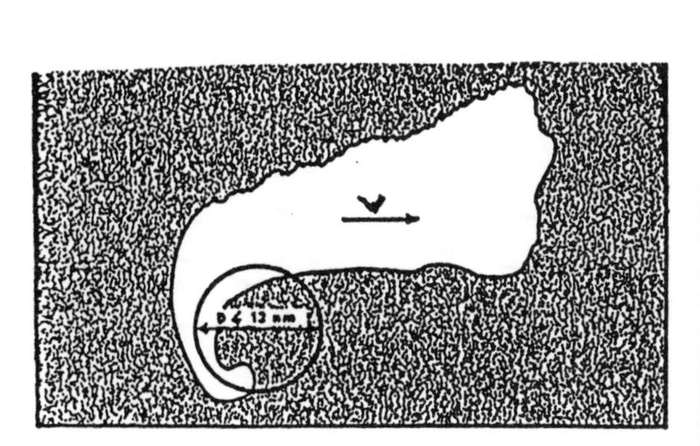

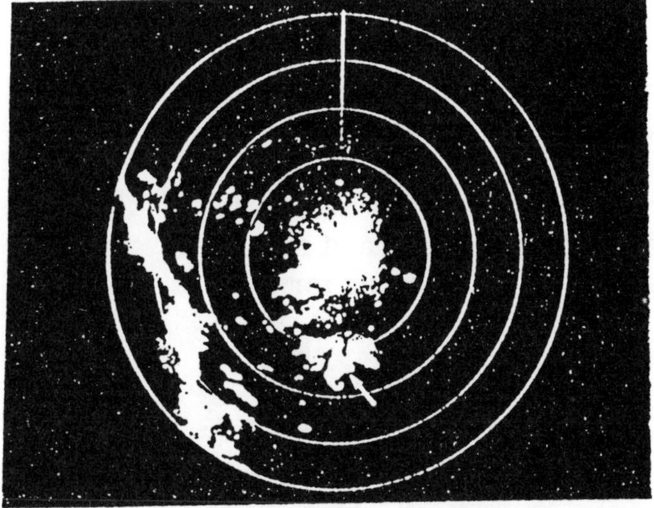

Figures 24-1. At left is a sketch of a HOOK ECHO, which usually comes of out the southwest part of the thunderstorm (in the Nrn. Hemisphere), and is typically a signature of a tornado. The figure at right is an actual radar display which at the arrow shows a thunderstorm with a hook echo. A tornado was on the ground where the hook was being observed. Each concentric ring is 25 nautical miles apart. (source: NWS)

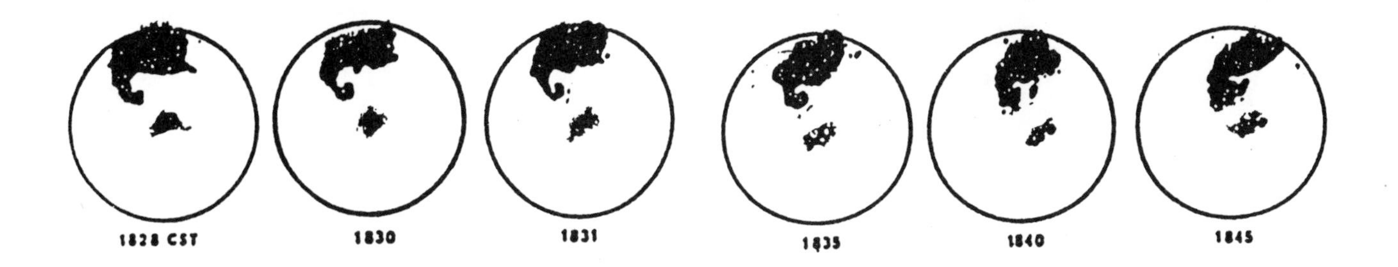

Figure 24-2. A time-series of radar images showing the development and movement of a tornado by its hook echo. This tornado passed just north of the radar site which is in the middle of the display. The numbers underneath each picture is the time in Central Standard Time. (source: NWS)

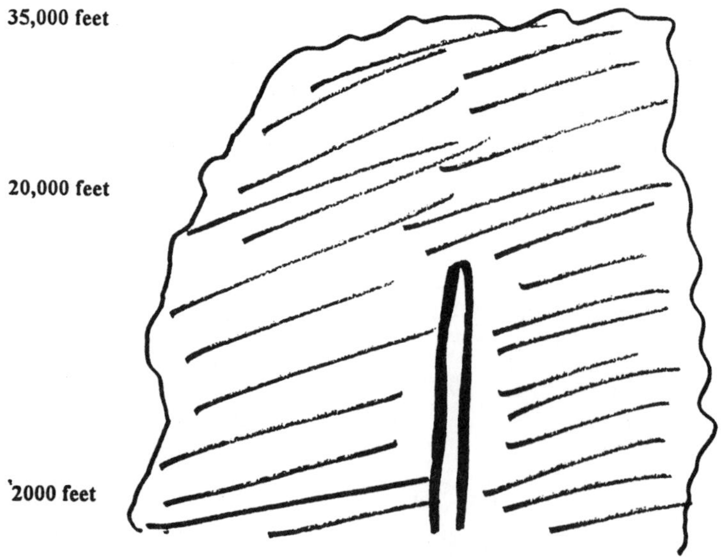

Figure 24-3. A bounded weak-echo region (BWER) on a radar's range/height display (RHI), which is when the rotation of the radar antenna is stopped so that it can scan up and down through a portion of the storm. A BWER is a region of no or very little precipitation which is surrounded by a radar echo of precipitation. It often indicates a region into which air is rapidly spiraling and rising and may be a tornado or developing tornado.

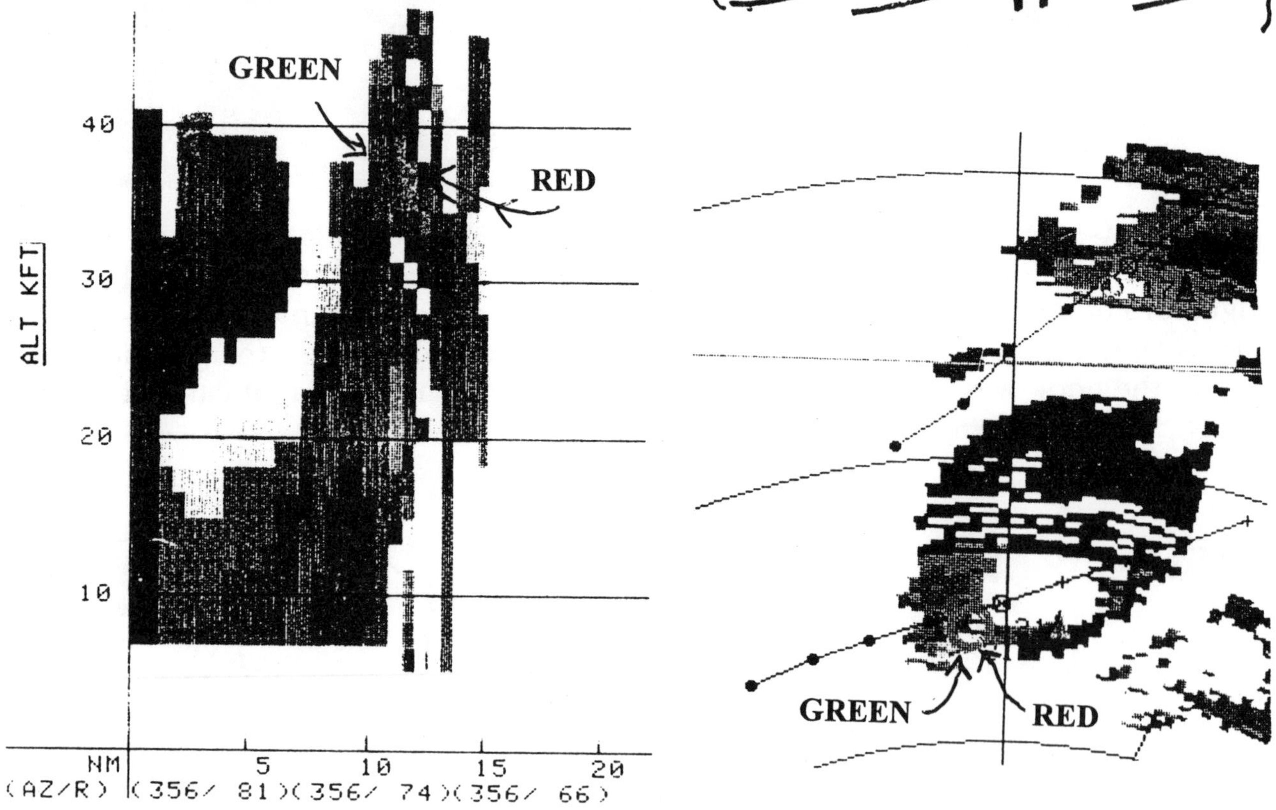

35,000 feet

20,000 feet

2000 feet

GREEN

RED

ALT KFT

40

30

20

10

NM 5 10 15 20

(AZ/R) (356/ 81)(356/ 74)(356/ 66)

GREEN

RED

Figures 24-4. A tornado on two Doppler radar wind displays. Both displays show the radial component of the wind heading either towards or away from the radar. The view at left is an RHI display within the cloud in one direction as the radar scans up and down the cloud, and the view at right is the wind display within the rotating radar beam. This is in color, and when the radial component of the wind shows a red area (wind heading away from the radar) right next to a green area (wind heading towards the radar), this means rotation is going on. These are signatures of a tornado in the Doppler wind velocity display. (source: NOAA)